Microplastics in Marine Ecosystem

This book addresses pertinent issues relating to microplastic pollution including its sources and sink of the microplastics and their environmental fate. It focuses on the impacts of microplastic pollution on marine life and human health. Available conventional methods and future solutions for the prevention and control of the marine microplastic pollution, such as bacterial and marine fungal biodegradation, membrane technology, and bioengineered microbes are included along with limitations and future challenges.

Features:

- Provides detailed insight into the marine microplastics pollution, fate, health impacts, and removal technology.
- Reviews ecological risks and environmental fate of microplastic pollution to the marine ecosystem.
- Describes control and prevention methods of the microplastics pollution.
- Covers global legislature for the mitigation of microplastic to the marine environment.
- Discusses the role of community participation for the reduction of microplastic emissions.

This book is aimed at researchers and professionals in environmental engineering, science, and chemistry, marine pollution, marine and aquatic science.

Microplastics in
Marine Ecosystem

Sources, Risks, Mitigation Technologies, and Challenges

Shobhika Parmar, Vijay Kumar Sharma
and Vir Singh

CRC Press
Taylor & Francis Group
Boca Raton London

CRC Press is an imprint of the
Taylor & Francis Group, an **informa** business

Designed cover image: © Shutterstock

First edition published 2023
by CRC Press
6000 Broken Sound Parkway NW, Suite 300, Boca Raton, FL 33487–2742

and by CRC Press
4 Park Square, Milton Park, Abingdon, Oxon, OX14 4RN

CRC Press is an imprint of Taylor & Francis Group, LLC

© 2023 Shobhika Parmar, Vijay Kumar Sharma and Vir Singh

ISBN: 978-1-032-31930-8 (hbk)
ISBN: 978-1-032-31932-2 (pbk)
ISBN: 978-1-003-31208-6 (ebk)

DOI: 10.1201/9781003312086

Typeset in Times
by Apex CoVantage, LLC

Contents

Contents

About the Authors

Shobhika Parmar earned a Ph.D. in Environment Science from G. B. Pant University of Agriculture & Technology, Pantnagar, India. She recently served as Postdoctoral Researcher in Kunming University of Science and Technology, Kunming, China. She also received the Yunnan Provincial Government Funding in China (2019). She has more than 11 years of research experience, focused on environmental issues and problems, remediation strategies. She has about 25 publications in journals of international repute and book chapters in edited books.

Vijay Kumar Sharma received his Ph.D. degree from Banaras Hindu University, Varanasi, India and completed two years of postdoctoral research in Kunming University of Science and Technology, Kunming, China. Currently, he is serving as a visiting scientist in Israel. He has more than 14 years of research experience. His research interest is in the bio-potential applications of endophytes and their mechanisms. He also has a keen interest in emerging environmental issues and mitigation, and works on few collaborative projects on the same area. Dr. Sharma has published more than 50 research articles and book chapters in international journals, and edited four books. He is also active as a reviewer and guest editor of some reputed journals. He has also actively participated in many national and international conferences, symposia, and workshops related to his research field in India, China, and Cyprus.

Vir Singh is currently serving as Emeritus Professor in the Department of Environmental Science, College of Basic Sciences and Humanities, G. B. Pant University of Agriculture and Technology, India. He has more than three decades experience in teaching, research, extension, project execution, and research supervision. He has been educated and trained in many universities and institutes in India and has also worked with many reputed organizations outside India, including International Centre for Integrated Mountain Development (ICIMOD) based in Kathmandu, Nepal; Galilee International Management Institute (GIMI) in Israel; and Friedrich-Schiller University in Germany. Prof. Vir Singh has published

55 books, many monographs, lab manuals, and more than 250 research papers, book chapters, and popular articles. His textbook *Environmental Plant Physiology* (Taylor and Francis, CRC Press, 2020) brings to the fore a botanical strategy for a climate-smart planet. He is also a Climate Reality Leader committed to creating awareness about the ongoing climate change and its long-term implications on every walk of life. He is also actively engaged in environmental writing. His articles on vital contemporary issues are being widely published in English and Hindi dailies and e-magazines.

Preface

The human species has almost completely transformed the biosphere into a semi-natural state, and finally seems to be assuming the shape of a "plastisphere". Plastics are ubiquitously present everywhere, in each of the Earth's spheres—the lithosphere, hydrosphere, atmosphere, and biosphere. Plastics in every shape and size, plastics in every color are in almost everything, in every walk of human life! And, of course, significant ill impact on human health is a self-revealing story of this relatively lesser understood pollution.

Constituting a large family of polymers, plastics are traditionally produced using fossil fuels and using about 4 to 8% of the fossil resources. With the range of plastic applications becoming wider and number of plastic users increasing constantly, consumption of fossil fuels for plastic production is likely to reach about 20% by 2050. Thus, dependent on scarce and non-renewable resources, conventional plastic production itself is based on the processes leading to environmental pollution through, for instance, excavation/digging out and refining of extra petroleum needed for plastic production and through production processes.

The most dramatic and worrisome scenario unfolding the formidable prospects of plastics is the extensively investigated presence of microplastics in the marine ecosystem. The microplastic pollution in the marine environment throughout the oceans, seas, and lakes of the globe is undoubtedly one of the most serious problems our contemporary world faces, and its mitigation is not so simple, and the global problem is not so easily manageable. Contamination of the planet's vastest environmental component—the hydrosphere—could be much more life-threatening than it has so far been perceived.

Already caught into an unprecedentedly grievous state of environmental disruptions and impending climate-linked disasters, our living planet is now haplessly trapped into a microplastic "web" that may be regarded as the third dimension of the planet's disaster. Burgeoning microplastic concentrations in the marine environments—that is, on most of the planet—and ubiquitous presence of the microplastics, including in the remotest uninhabited Antarctica and the Arctic region, are not only creating ecological and human health-related problems as could be expected from yet another type of a pollutant, but this pollution is all set to significantly contribute to fortifying the processes leading to climate change.

Microplastics in Marine Ecosystem: Sources, Risks, Mitigation Technologies, and Challenges offers a new in-depth subject matter covering all aspects relating to the escalating microplastic pollution with special focus on global marine ecosystems, inviting the students, teachers, and researchers from diverse disciplines to enhance their knowledge base, and attracting the attention of planners, policy makers, environmentalists, ecologists, social workers, and all those committed to the cause of the planet Earth to get a wide exposure to the various sources, risks, mitigation strategies, and challenges pertaining to the microplastic pollution in the marine ecosystem.

The book comprises 11 chapters explaining in detail the latest research-based supporting information about the microplastics in the marine ecosystem: their various sources and sinks, analytical methods for their identification and assessments, ecological risks and environmental fates, human health impacts, biodegradation by microbes and fungi, various membrane technologies used for their removal, bioengineering advances in plastic biodegradation, biopolymers as an alternative to conventional plastics, global legislature for microplastic mitigation, community participation for the reduction in microplastic emissions, and recent cutting-age solutions to prevent microplastic pollution.

Microplastics in the Earth's hydrosphere—the planet's largest environmental component—is emerging as one of the most pressing environmental issues. Many disciplines are dealing with this deepening environmental crisis. *Microplastics in Marine Ecosystem* provides comprehensive material along with vital technological/biotechnological applications instrumental in effectively managing microplastic pollution and resolving terrible issues bubbling up in our contemporary world. The book is of great worth for undergraduate, graduate, postgraduate, and research students as well as for their teachers and for scientists working in environmental science and allied areas. The book is also an exemplary guide for professionals, policy makers, planners, industrialists, environmentalists, environmental engineers, biotechnologists, managers in the plastic industry, and many officials and workers in United Nations organizations, governments, and non-government organizations.

Shobhika Parmar

Vijay Kumar Sharma

Vir Singh

Abbreviations and Acronyms

μg	Microgram
μm	Micrometer
AAB	Acetic acid bacteria
ABS	Acrylonitrile butadiene styrene
ADH	Attention Deficit Hyperactivity
Al	Aluminum
ARB	Antimicrobial resistance bacteria
ARGs	Antibiotic resistance genes
As	Arsenic
ASTM	American Society for Testing and Materials
ATR	Attenuated total reflection
Ba	Barium
BIS	Bureau of Indian Standards
BOD	Biochemical oxygen demand
BPA	Bisphenol A
Br	Bromine
bw	bodyweight
Caco-2	Cancer coli "colon cancer"
CBD	Convention on Biological Diversity
Cd	Cadmium
cm	Centimeter
CMCase	Carboxymethyl cellulase
CMS	Conservation of Migratory Species
Co	Cobalt
CO_2e	Carbon dioxide equivalent
CONTAM	Contaminants in the Food Chain
CPCB	Central Pollution Control Board
Cr	Chromium
CSWRCB	California State Water Resources Control Board
Cu	Copper
DAF	Dissolved air floatation
DEHP	Diethylhexyl phthalates
DM	Dynamic membrane
DNA	Deoxyribonucleic acid
ECETOC	European Centre for Ecotoxicology and Toxicology of Chemicals
ECHA	European Chemical Agency
EDCs	Endocrine-disrupting chemicals
EDS	Energy dispersive spectroscopy
EFSA	European Food Safety Authority
EIA	Environmental impact assessment
EPA	Environmental Protection Agency
EPR	Extended Producer Responsibility

EPS	Exopolysaccharides
EPS	Expanded polystyrene
EPS	Extracellular polymeric substances
EU	European Union
FAO	Food and Agriculture Organization
FPA	Focal plane array
FRED	Floating Robot for Eliminating Debris
FTC	Federal Trade Commission
FTIR	Fourier transform infrared
GAC	Granular activated carbon
GESAMP	Group of Experts on the Scientific Aspects of Marine Environmental Protection
GHG	Greenhouse gas
GPGP	Great Pacific Garbage Patch
GPWM	Global Partnership on Waste Management
Gt	Gigaton
HA	Hyaluronic acid
HABs	Harmful algal blooms
hASCs	Human adipose stem cells
HDPE	High-density polyethylene
Hg	Mercury
HRT	Hydraulic retention time
IMO	International Maritime Organization
IR	Infrared
IWMS	Integrated Waste Management System
kg-bw	Kilogram-bodyweight
LAB	Lactic acid bacteria
LCA	Life cycle assessment
LDPE	Low-density polyethylene
LLDPE	Linear low-density polyethylene
LMCO	Laccase-like multicopper oxidase
LMWPE	Low molecular weight polyethylene
MARPOL	International Convention for the Prevention of Marine Pollution
MBR	Membrane bioreactor
mg	Milligram
MHET	mono(2-hydroxyethyl) terephthalic acid
ml	Milliliter
MLSS	Mixed liquor suspended solids
mm	Millimeters
MMT	Million metric tons
Mn	Manganese
MP	Microplastics
MRGs	Metal resistance genes
MSFD	Marine Strategy Framework Directive
MSW	Municipal solid waste

nano-TA	Nano-thermal analysis
ng	Nanogram
NGO	Non-government organization
NOAA	National Oceanic and Atmospheric Administration
NP	Nanoplastics
O_3	Ozone
P(3HB-co-3MP)	Poly(3-hydroxybutyrate-co3-mercaptopropionate)
PA	Polyamide
PAC	Polyaluminum chloride
PAEs	Phthalate esters
PAHs	Polycyclic aromatic hydrocarbons
PAM	Polyacrylamide
Pb	Lead
PBA	Polybutyric adipate
PBAT	Poly(butylene adipate-co-terephthalate)
PBDEs	Polybrominated diphenyl ethers
PBS	Polybutylene succinate
PBT	Polybutylene terephthalate
PCA	Protocatechuic acid
PCBs	Polychlorinated biphenyls
PCL	Polycaprolactone
PCL	Poly(Ɛ-caprolactone)
PE	Polyethylene
PEA	Polyethylene adipate
PEG	Polyethylene glycol
PES	Polyethylene succinate
PET	Polyethylene terephthalate
PEVA	Polyethylene-vinyl acetate
PGA	Poly-glutamic acid
PHA	Polyhydroxyalkanoate
PHB	Polyhydroxybutyrate
PHBV	Poly(3-hydroxybutyrate-co-3-hydroxyvalerate)
PHV	Polyhydroxyvalerate
PLA	Polylactic acid
PLC	Polymers of low concern
PMMA	Poly(methyl methacrylate)
POPs	Persistent organic pollutants
PP	Polypropylene
PPA	Polypropylene adipate
PS	Polystyrene
PU/PUR	Polyurethane
PVA	Polyvinyl alcohol
PVC	Polyvinyl chloride
Pyr-GC-MS	Pyrolysis gas chromatography/mass spectrometry
R&D	Research and development

REACH	European Union Law for Registration, Evaluation, Authorization and Restriction of Chemicals
RO	Reverse osmosis
ROS	Reactive oxygen species
SAPEA	Science Advice for Policy by European Academies
Sb	Antimony
SBSTTA	Subsidiary Body on Scientific, Technical and Technological Advice
SDG	Sustainable Development Goal
SEA	Strategic environmental assessment
SEM	Scanning electron microscope
SMM	Sustainable Material Management
Sn	Tin
SPCB	State Pollution Control Board
SRT	Sludge retention time
SUP	Single-use plastic
TCA	Tricarboxylic acid
TEM	Transmission electron microscopy
Ti	Titanium
TPS	Thermoplastic starch
UF	Ultrafiltration
ULB	Urban Local Body
UN	United Nations
UNCLOS	United Nations Convention on the Law of the Sea
UNEP	United Nations Environment Program
USA	United States of America
USEPA	Environmental Protection Agency of USA
UV	Ultraviolet
VC	Vinyl chloride
viSNE	visual stochastic network embedding
WCED	World Commission on Environment and Development
WFS	World Food Summit
WoSCC	Web of Science Core Collection
WWTP	Wastewater treatment plant
XRF	X-ray fluorescence
Zn	Zinc

1 Sources and Sinks of Microplastics

1.1 WHAT ARE MICROPLASTICS?

Microplastics are tiny or microscopic particles of plastics smaller than 5 mm. Microplastics are the most frequently detected plastics worldwide that have been identified as a widespread global threat to human health and the environment. Plastics are an indispensable part of our daily lives, revealing their ubiquitous presence. Plastics form parts of our clothing to our mobile phones to our electronic appliances to even our food packaging. Although plastic was first invented in the early 1900s, its popularity, use and large-scale production increased dramatically in the 1950s. Since then, until the last decade, more than 8000 million metric tons (MMT) of new plastic has been produced globally. According to an estimate, over 6300 MMT of plastic garbage was generated by 2015, with approximately 9% being recycled, 12% being burnt, and 79% ending up in the landfills or the environment (Geyer et al. 2017). Moreover, by 2050 there will be 12,000 MMT tons of plastic garbage in the landfills and the environment.

Larger particles of plastic items and waste break down into smaller and smaller fragments. These small plastic fragments are called "microplastics". The term microplastic was first proposed by Thompson et al. in 2004. The defined size range of microplastics is between 1–5000 µm. In addition, sometimes deliberately very tiny pieces of plastics are manufactured such as microbeads that have applications in beauty and personal care products like toothpastes and cleansers. These microplastics easily pass unchanged through waterways, ending up in the oceans. Microbeads were banned in the United States in 2015. Nevertheless, this was only a very small part of the problem. In fact, microplastics are a huge problem worldwide. Every year, around 4.8 to 12.7 MMT of plastic debris enters the ocean from the plastic waste produced by the 192 coastal countries across the globe (Jambeck et al. 2015); further, about 1.15 to 2.41 MMT of plastic debris enters the ocean through rivers (Lebreton et al. 2017). Fragmentation of this plastics debris generates microplastics that pollute marine habitats including beaches, lagoons, mangrove forests, and deep oceans. Microplastics contamination was listed the second most serious scientific problem in the realm of environmental and ecological science research at the 2015 UN Environment Conference.

Microplastics are ubiquitously observed in the marine environment, and have been found in every system examined. Microplastics have been detected even at the farthest locations such as the Arctic region and in deep-sea sediment. These particles are frequently mistaken for prey by marine organisms because of their lightweight, omnipresent, long-lasting nature and wide range of size and colors. Thus, it enters into the food web and even reaches the human diet through it. Microplastics are toxic

DOI: 10.1201/9781003312086-1

to a wide range of marine life. Microplastics have toxicological effects on human health, including oxidative stress and cytotoxic impacts on the human body when it is exposed to microplastics or consumes contaminated food. Recently, for the first time microplastics have been detected in human blood, with these tiny particles found in almost 80% of the tested people. Once released into the environment, microplastics become an inseparable part of marine ecosystems. There is still no concrete evidence about the half-life of the microplastics, but it is commonly believed among the scientific community that they can withstand varying environmental conditions, resist environmental degradation, and will last long in the environment after discharge.

Since microplastics originate from plastics, they can be made of different polymers such as polypropylene, polyethylene, polystyrene, polyethylene terephthalate, polyether urethane, low-density polyethylene, polyacrylates, alkyd resin, polyamide, acrylamide, polyvinyl chloride, polyether sulfone, polyvinyl alcohol, etc. Marine microplastics debris is a complex cocktail of chemicals and contaminants. These include mainly the monomers and polymers of which they are primarily made or the by-products of the manufacturing process and contaminants present in the environment. Often other contaminants such as antibiotics, potentially toxic metals, and organic pollutants interact with the microplastics, readily accumulate and concentrate at a very high level. Furthermore, microplastics leach hazardous substances and chemical additives, including phthalates, formaldehyde, bisphenol A, etc.

1.1.1 DEFINITIONS

- The term microplastics was first coined to describe the accumulation of microscopic plastics particles in marine sediments and in the water column of European waters (Thompson et al. 2004).
- Ng and Obbard (2006) defined the formation of microplastics: "Plastics at sea eventually undergo fragmentation, leading to the formation of microscopic particulates of plastic called microplastics".
- Arthur et al. (2009) revised the initial term by proposing an upper size limit; thus microplastics are the "plastic particles smaller than 5 mm", irrespective of their origin.
- Based on the origin, Cole (2011) classified microplastics as primary microplastics (manufactured to be microscopic size) or secondary microplastics (resulting from degradation and fragmentation of the larger plastics). It was stressed that there is a key requirement to use a clear and uniform definition, as well as additional sizes such as nano- and mesoplastics.
- GESAMP (Group of Experts on the Scientific Aspects of Marine Environmental Protection) defined microplastics as plastic "particles in the size range of 1 nm to less than 5 mm" (GESAMP 2015).
- The EFSA (European Food Safety Authority) Panel on Contaminants in the Food Chain (CONTAM) (2016) also stated that no internationally recognized definition is available. They defined microplastics as a "heterogeneous mixture of differently shaped materials referred to as fragments, fibres, spheroids, granules, pellets, flakes or beads, in the range of 0.1 to 5,000 μm". Nanoplastics are the plastics particles of nanoscale dimensions (0.001 to 0.1 μm).

- Frias and Nash (2019) cited that there is need to find a consensus on the way microplastics should be defined, which can be applied for legislative purposes, research, and reporting. They proposed the following definition: "Microplastics are any synthetic solid particle or polymeric matrix, with regular or irregular shape and with size ranging from 1 μm to 5 mm, of either primary or secondary manufacturing origin, which are insoluble in water".
- In 2020, CSWRCB (California State Water Resources Control Board) proposed a definition of microplastics in drinking water as: "solid polymeric materials to which chemical additives or other substances may have been added, which are particles which have at least three dimensions that are greater than 1 nm and less than 5,000 μm. Polymers that are derived in nature that have not been chemically modified (other than by hydrolysis) are excluded". (Coffin 2020)

1.1.2 DISTRIBUTION AND COMPOSITION

In the marine environment plastics are the major (60–80%) solid debris (Yaranal et al. 2021). Distribution of microplastics has been evaluated in many regions and marine creatures. However, the number of studies is limited globally, so complete information on the distribution of microplastics is still not known. The breakdown of larger plastic through chemical, mechanical, and biological degradation in the environment generates secondary microplastics, which are largely composed of a parent polymer. Polyvinyl chloride, polyethylene, polystyrene, polypropylene, nylon, polyethylene terephthalate, and acrylonitrile butadiene styrene are some of the common secondary microplastics. While the primary microplastics are produced, their composition varies depending on the application (cosmetic, personal care product, drugs, cleansers, or other industrial uses). For example, polyethylene is the most common microplastic present in cosmetics; paints have polyurethane, acrylic, epoxy, alkyd, or chlorinated rubber binders. Microplastics in the marine system may consist of beads, fibers, films, fragments, pellets, and Styrofoam. These microplastics can get into the marine environment via different modes like maritime activities, freshwater ways, surface runoff from the terrestrial microplastics, or directly through wind. There is lot of heterogeneity in the occurrence and distribution of microplastics in different regions (Table 1.1). Their spatial distribution is very uneven. Anthropogenic activities like wastewater and industrial discharge, mariculture, coastal settlements, and hydrodynamics have an influence on the distribution. Microplastics having density more than seawater accumulate at the bottom in marine sediments due to gravity while the microplastics having density less than that of seawater float over the surface water and subsequently pollute the coastline as it flows back due to waves. In the marine environment microplastics have been detected globally from different and far-reaching locations such as Arctic regions, different coastal areas, river estuaries, etc.

1.1.3 CLASSIFICATION OF MICROPLASTICS

For the sake of convenience and clarity, microplastics are classified into subtypes based on the size or shape, whether they were manufactured as microplastics or formed because of physicochemical degradation.

TABLE 1.1

Distribution of microplastics in different regions studied across the globe.

Location	Size range	Abundance	Particle type	Polymer component	Reference
Belgian marine sediments	38 μm–1 mm	390 particles kg^{-1} dry sediment	Fibers, granules, films, polystyrene spheres	Polypropylene, nylon and polyvinyl alcohol, nylon, polystyrene	Claessens et al. (2011)
Incheon/ Kyeonggi Coastal Region	50 μm–5 mm	48.0–360 (152±92.2) particles m^{-2}.	n.a.	Polypropylene, polyethylene, polyvinylchloride, polyvinyl alcohol, polyvinyl sulfate, polyethylene terephthalate, expanded polystyrene, and paint particles (alkyd and poly[acrylate/ styrene])	Chae et al. (2015)
Urban estuaries of China (Jiaojiang, Oujiang, Minjiang)	90% <5 mm	680.0 ± 284.6 to 1245.8 ± 531.5 particles m^{-3}	Fibers, granules and films	Polypropylene and polyethylene	Zhao et al. (2015)
Atlantic Ocean	Synthetic polymers were <5 mm in length	0 to 8.5 particles m^{-3}	Majority of microplastics were fibers and a few fragments while Rayon particles were only fibers	37% were synthetic polymers and 63% as Rayon. Polymer type include polyester, polyamide blends, or acrylic/ polyester blends, polyamide, polypropylene, acrylic, polyvinyl chloride, polystyrene, and polyurethane	La Daana et al. (2017)
Bohai Sea, China	Microplastics (0.3–5 mm) were 55% of the total particles	0.33 ± 0.34 particles m^{-3}	Lines, films, fragments, spheres, fibers, pellets, and beads	Polyethylene, polypropylene, polystyrene, and polyethylene terephthalate were the main polymer	Zhang et al. (2017)

Location	Size range	Abundance	Particle type	Polymer component	Reference
Beaches of Lima, Peru	54.2% were from 1 to 2.8 mm in size, the remaining 45.8% were from 2.9 to 4.75 mm.	16.67 ± 4.26 to 489.7 ± 143.5 particles m^{-2}	78.3% were foams, 17.38% were fragments	Polyethylene, polypropylene, polystyrene, polypropylene	De-la-Torre et al. (2020)
Kuwait coastal areas	n.a.	n.a.	Filaments, fragments, films	Polypropylene, polyethylene, polystyrene.	Saeed et al. (2020)
Salina Island, Italy	0.1 to 2.0 mm	Sediments: 49.0 ± 1.4 items kg^{-1} (landslides), 153.5 ± 41.7 items kg^{-1} (cliff), 106.0 ± 104.7 items kg^{-1} (banks)	n.a.	Polyvinylchloride, polyethylene, polypropylene, polyamide, polystyrene, nylon	Renzi et al. (2020)
Northern Tyrrhenian Sea	(1–5 mm) accounted for 84.5% of particles	163.2 ± 77.4 to 331.9 ± 161.6 particles m^{-2}	Filaments (73.8%), fragments (17.2%), films (9.0%)	Polyethylene terephthalate (5.8%), polyethylene (41%), thermoplastic polyurethane (34.1%), acrylonitrile-butadiene-styrene copolymer (1.9%), nylon (44.7%), epoxy resin (2.4%)	Mistri et al. (2020)
Klang River estuary, Malaysia	0.3 to 1 mm	0.5 to 4.5 particles L^{-1}	Fibers, fragments, and pellets	Polyamide and polyethylene were the main polymer	Zaki et al. (2021)
White Sea basin (Russian Arctic)	Most of the particles were <0.33 mm	0.19 to 1.0 particles m^{-3}.	Fragments, films, and fibers	Polyethylene, polyethylene terephthalate, polyvinylchloride	Ershova et al. (2021)
Marina beach, India	0.1 to 3.0 mm	0.1	Fiber (15%), foam (12%), pellet (6%), film (10%), fragment (33%), and unidentified particles (24%)	Polyethylene (29.9%), polypropylene (16.9%), polyvinylchloride (10.9%), PET (9.9%), and nylon (7.9%)	Venkatramanan et al. (2022)

(Continued)

TABLE 1.1

(Continued)

Location	Size range	Abundance	Particle type	Polymer component	Reference
European Arctic (surface sediments)	0.3–2 mm (81.82% to 81.91%)	Average 721.42 ± 217.89 to 783 ± 530.28 pieces kg^{-1}	Fibers (91.59% to 97.10%)	Polyethylene and polypropylene were the most common polymers followed by polyvinylchloride and nitrile	Choudhary et al. (2022)
Marine sediments of Chilean fjords	<5 mm	Average 166.7 ± 92.1 items kg dw^{-1}	Fibers (88%), fragments (10%), films (2%)	Polyethylene terephthalate and acrylics, polypropylene, polyurethane	Jorquera et al. (2022)
Beaches along southeast coast of India	<5 mm	370 to 708 particles kg^{-1} dry weight	Filaments > films > fragments	Polyethylene, polypropylene	Kaviarasan et al. (2022)
Beaches along the coast of Qingdao, China	0.1–1 mm	Average 91.11 ± 26.76 to 147.78 ± 34.80 items m^{-2}	Lines (80.5%) and fragments (7.9%)	Rayon (41.8%) and polyethylene terephthalate (16.9%) were the main polymer	Gao et al. (2021)

Note: n.a.—not available.

1.1.3.1 Classification according to Size

There is lot of variation in the literature in the categorization of the plastic particles based on the size (Figure 1.1). The widely accepted defined size range of microplastics is between 1–5000 μm. Further, microplastics are divided into small (1–1000 μm) and large (1–5 mm). Larger plastics were classified into megaplastics (those with a diameter more than 1000 mm), macroplastics (those with a diameter between 250 to 1000 mm), and mesoplastics (those with a diameter between 5 to 250 mm) (Menéndez-Pedriza and Jaumot 2020). Microplastics below the 1 μm size should be referred to as nanoplastics, a relatively less studied and unknown component of the marine litter. However, these four size categories (macroplastic, mesoplastic, large microplastic, small microplastic) are too broad. To study the impact on the environment and elutriation process, it is essential to determine the particle size ranges. Kedzierski et al. (2016) proposed a new granulometric approach for the classification of different plastic particle-size fractions in the marine sediments (Table 1.2). This system was established using Gradistat sedimentological classification that takes into account existing classifications (Galgani et al. 2013; Van Cauwenberghe et al. 2015). Hartmann et al. (2019) identified the main defining criteria and classifiers that should be adopted for defining and categorizing plastic debris (Table 1.3). Chemical composition was the integral part of the criterion that categorizes based on whether the particles are synthetic, heavily or slightly modified, comprised of additives, copolymers,

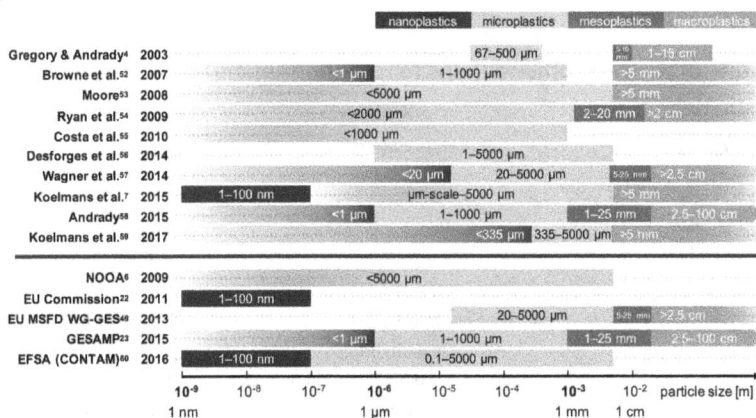

FIGURE 1.1 Variation in the size-based categorization of plastic particles according to size as described in scientific and institutional literature.

Source: Adapted from Hartmann et al. (2019).

TABLE 1.2
Size-based particle classifications for plastic. This classification was based on Gradistat classification.

Size (mm)	Plastic (after Kedzierski et al., 2016)		Plastic (after Van Cauwenberghe et al., 2015)
	Categories	Sub-categories	
1024		Very coarse (Ma$_{vc}$)	
512	Macroplastic (Ma)	Coarse (Ma$_c$)	
256		Medium (Ma$_m$)	Macroplastics
128		Small (Ma$_s$)	
64		Very small (Ma$_{vs}$)	
32	Mesoplastic (Me)	Very coarse (Me$_{vc}$)	
16		Coarse (Me$_c$)	25
8		Medium (Me$_m$)	Mesoplastic
4		Fine (Me$_f$) $-\frac{r2}{r1}-$	5
2		Very Fine (Me$_{vf}$)	Large microplastic
1	Microplastic (Large; MI^L)	Crossover (MI$^L_{co}$)	1
0.5		Coarse (MI^L_c)	
0.25		Medium (MI^L_m)	
0.125		Fine (MI^L_f)	
0.063		Very Fine (MI$^L_{vf}$)	
0.031	Microplastic (Small; MI^S)	Very coarse (MI$^S_{vc}$)	Small microplastic
0.016		Coarse (MI^S_c)	
0.008		Medium (MI^S_m)	
0.004		Fine (MI^S_f)	
0.002		Very Fine (MI$^S_{vf}$)	
0.001	Nanoplastic (Na)	Crossover (Na$_o$)	0.001 Nanoplastic

Source: Kedzierski et al. (2016; Blott and Pye 2001).

TABLE 1.3

Overview of the recommendations for a definition and classification of plastic debris.

	Criterion	Recommendation	Examples
I	Chemical Composition		
I(a)	Polymers	All synthetic polymers:	
	☑ Include	➢ Thermoplastics	All commodity plastics
		➢ Thermosets	Polyurethanes, melamine
		➢ Elastomers	Synthetic rubber
		➢ Inorganic/hybrid	Silicone
	☑ Include	Heavily modified natural polymers (semi-synthetic)	Vulcanized natural rubber, regenerated cellulose
	☒ Exclude	Slightly modified natural polymers	Dyed natural fibers
I(b)	Additives		
	☑ Include	All polymers included in I(a) disregarding their additive content	Plasticized PVC with >50% additives
I(c)	Copolymers		
	☑ Include	All copolymers	ABS, EVA, SBR
I(d)	Composites		
	☑ Include	All surface coatings containing polymers as essential ingredient	Reinforced polyester and epoxy
	☑ Include	All surface coatings containing polymers as essential ingredient	Paints containing polyester, PUR, alkyd, acrylic, epoxy resin
	☑ Include	Tire wear (and road) particles	–
	[?] Open question	Is it necessary to define a minimum polymer content?	
II	**Solid state**		
	☑ Include	All polymers with a T_m or T_g >20°C	Examples same as I(a)
	☒ Exclude	Polymer gels	PVA, PEG
	[?] Open question	Should wax-like polymers (T_g, <20°C) be included?	
III	**Solubility**		
	☑ Include	All polymers with a solubility <1 mg L^{-1} at 20°C	Examples same as I(a)
IV	**Size**		
		➢ Nanoplastics: 1 to <1000 nm	
		➢ Microplastics: 1 to <1000 μm	
		➢ Mesoplastics: 1 to <10 mm	
		➢ Macroplastics: 1 cm and larger	

The largest dimension of the object determines the category. Comprehensive reporting of multiple dimensions is preferred (e.g., for fibers).

V	**Shape and structure**		
		➢ Spheres: Every surface point has the same distance from the center	
		➢ Spheroid: Imperfect but approximate sphere	
		➢ Cylindrical pellet: Rod-shaped, cylindrical object	

Criterion	Recommendation	Examples
		➢ Fragment: Particle with irregular shape
		➢ Film: Planar, considerably smaller in one than in the other dimensions
		➢ Fiber: Significantly longer than wide in two dimensions
	Additional information on the structure (e.g., porosity) can be included.	
VI	**Color**	
		Not crucial but useful in some biological contexts. Use a standardized color palette.
VII	**Origin** (Optional)	
		Primary: Intentionally produced in a certain size
		Secondary: Formed by fragmentation in the environment or during use
	Origin should only be used if the primary origin can be established.	

Source: Adapted from Hartmann et al. (2019).

and composites. Another criterion takes account of the natural state (i.e., if present as solid and insoluble at 20°C). Additionally, criteria such as size, shape, color, and origin were also recommended.

Microplastics of different shapes such as lines, films, fragments, spheres, fibers, pellets, granules, foams, and beads were identified from various marine samples from different regions (Table 1.1). Sometimes color (white, yellow, green, blue, red, grey, black, transparent, etc.) is also included as a classifier to study microplastics (Jorquera et al. 2022). It is not easy to draw any conclusion about the origin and chemistry of the microplastics just by identifying shape and color. However, including color as an extra descriptor can be important from the biological perspective. Colored microplastics are easily mistaken by the marine animals as their prey.

1.1.3.2   Nanoplastics

Microplastics can subsequently degrade into nano-sized fragments. Due to their relatively very small size, these nano-sized fragments can easily penetrate and accumulate in organs. Therefore, their impact on the environment will be totally different as compared to those imparted by microplastics. The nano-sized fragments of plastics are termed as nanoplastics. At present, there is no definite consensus on the size of nanoplastics; however, widely adapted size range is between 1 to 1000 nm (Kedzierski et al. 2016; Hartmann et al. 2019). Largely nanoplastics originate by the degradation and fragmentation of microplastics and larger plastic debris, thus, are highly polydisperse and heterogeneous.

1.1.3.3   Classification according to Origin

As discussed, according to their origin microplastics can be divided into primary and secondary types.

- Primary microplastics include cosmetic and personal care products, drugs, cleansers, paints, gels, etc. They are usually manufactured as microbeads, and are purposely added to products for application. Further unintentionally released microplastics such as pelleted raw materials are also included in primary category.

- Secondary microplastics result from the fragmentation of bigger plastics through chemical, mechanical, and biological degradation. For example, the plastic bottles and bags, disposable plastics cups, and other packaging materials that people throw here and there ultimately turn into secondary microplastics.
- Another class of microplastics is in between primary and secondary, while we include them in the primary category. They are not manufactured to micro-sizes (for instance, microbeads and abrasive scrubbers), nor are they formed due to fragmentation of plastic (for example, fibers from synthetic fabrics, tire dust, road markings, etc.).

1.2 WHAT ARE THE SIGNIFICANT EMISSION PATHWAYS OF MICROPLASTICS INTO THE ENVIRONMENT?

Worldwide only a fraction (1%) of the plastic waste enters directly into the marine environment. Most of the emissions are to the terrestrial land and freshwaters either as primary or secondary. Sea-based sectors like marine aquaculture, marine and coastal tourism, seabed mining, shipping and offshore industries release plastics directly into the marine environment. However, as compared to the marine-derived emissions, most of the plastics enter indirectly into the oceans through the emissions on the land. Jambeck et al. (2015) made an estimate of the land-based plastic waste generation (within the range of 50 km from the coast) along the 192 coastal nations of the world, which was evaluated to be 275 MMT for the year 2010. Of this, 4.8–12.7 MMT was projected to enter oceans. For any country, the size of the population and the percentage of the waste managed can tell us a lot about the plastics waste generated by the country and its plastics contribution to the marine environment. For example, the highest waste-generating countries like China, Indonesia, Philippines, Vietnam, Sri Lanka, Thailand, Egypt, Malaysia, Nigeria, and Bangladesh have some of the largest coastal populations and lack proper waste management infrastructure (Jambeck et al. 2015).

In later years, two important studies calculated the worldwide plastics emissions into the oceans through the rivers. The first study took account of the plastic load of 40,760 catchments across the world and estimated that every year about 1.15 to 2.41 MMT ends up in the ocean from rivers (Figure 1.2) (Lebreton et al. 2017). Moreover, 67% of this annual load came from the top 20 most polluted rivers, which are mostly located in Asia. A more recent study that considered 100,887 stream and river outlets in their model, out of which 31,904 locations discharged plastic into the ocean, investigated about 0.8–2.7 MMT being released into the oceans in 2015 (Meijer et al. 2021). Further, more than 1000 rivers were responsible for 80% of total annual global emissions. The freshwater ways are the common link between land microplastics and marine microplastics. The surface runoff takes a part of the terrestrial microplastics into the channels and drains through which the microplastics enter into the river if not treated or removed. The rivers transmit microplastics to the marine environment. Sometime particles from the river may flow back to the land through flooding. Lighter terrestrial microplastics are directly sent to rivers and the ocean by the wind. Several microplastics emission pathways have been identified (Table 1.4).

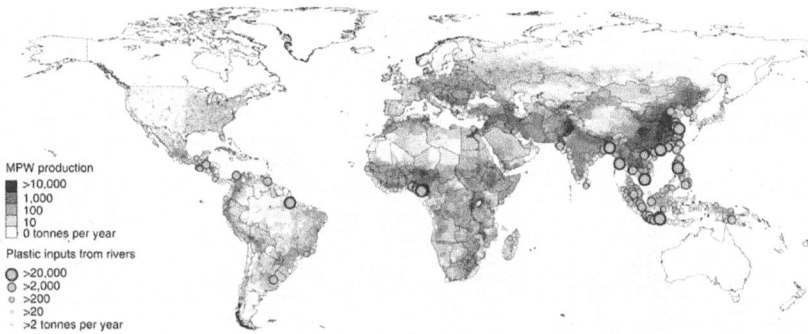

FIGURE 1.2 Plastics inputs from the rivers into oceans in tons per year.

Source: Adapted from Lebreton et al. (2017).

Any activity that generates microplastics or nanoplastics can be characterized as point source or diffuse source based on the location and time and with reference to the primary or secondary characteristics of the particles released (Allen et al. 2022). Point source refers to the microplastics emission at a specific time and particular location (e.g., emission from wastewater treatment to a freshwater body, emission from a recycling plant, or emission from a plastic factory). Diffuse source refers to microplastics emission having no definite time and location (e.g., tire abrasions and wearing-off road paints). Some of the identified sources of marine microplastics pollution include pellets and fragments used in plastic manufacturing, fabricating, and recycling; improper disposal of solid waste, sewage, and wastewater; plastics used in the construction industries; pellets, tire dust, and road paint abrasions generated by vehicles; consumer goods, packaging, and microbeads from the tourism industries. Besides, containers, plastic bags, bottles, caps, plastic disposable cutlery, and other items used for food and drinks packaging; textile fibers from individual consumers and the textile industries; sports, fisheries, aquaculture, agriculture, shipping/offshore industries are also sources of microplastics (Table 1.4). Further, some latest studies using data models determined that plastics' presence in the atmosphere (together with nanoplastics) is an important part of the marine plastic cycle (Allen et al. 2022). Every year, between 0.013 and 25 MMT microplastics (+nanoplastics) are potentially transmitted into the marine atmosphere and released into the oceans (Allen et al. 2022).

Mechanical plastic waste recycling and wastewater effluents are the recognized point source pathway of microplastics pollution entering the aquatic environment (Suzuki et al. 2022; Ziajahromi et al. 2017). Basic acts in our daily life, such as cutting plastics with scissors or knives, tearing it with our hands, or just simply twisting to open plastic bags/containers/tapes/caps, can produce microplastics of 0.46 to 250 microplastics per cm (Sobhani et al. 2020). Polyester and other textiles also generate microplastics, on an average of 0.025 mg microplastics fibers per gm per wash (Hernandez et al. 2017). Indeed, analysis of wastewater samples from household washing machines showed that more than 1900 fibers are produced per wash from single garment (Browne et al. 2011). A study in 2021 found that 70% of the personal

TABLE 1.4
Microplastics emission pathways into the environment.

Category		Source	Type	Entry point
Producers/ Transformers		Plastic factories, fabricators and recyclers	Pellets and fragments	Rivers, coastline, atmosphere
		Solid waste	Poor disposal of waste	Rivers, coastline, atmosphere
Waste management		Electronics waste	Plastics flakes generated while dismantling and milling of electronic waste	Rivers, coastline
		Sewage and wastewater	Microbeads, fragments, fibers	Rivers, coastline
		Construction	Expanded polystyrene, packaging, fiber-reinforced polymer, polycarbonate, polyethylene, polypropylene, polyvinyl chloride, acrylic, composites	Rivers, coastline, atmosphere

Category		Source	Type	Entry point
		Land transport	Pellets, tire dust, road paint abrasions	Rivers, coastline, atmosphere
		Land and tourism	Consumer goods, packaging, microbeads, textile fibers	Rivers, coastline, atmosphere
Sectoral consumers		Marine tourism	Consumer goods, packaging, microbeads, textile fibers	Marine
		Textile industries	Textile fibers	Rivers, coastline, atmosphere
		Sports	Synthetic turf, fibers	Rivers, coastline, atmosphere
		Fisheries	Fishing gear, packaging	Rivers, coastline, marine

(Continued)

TABLE 1.4
(Continued)

Category		Source	Type	Entry point
	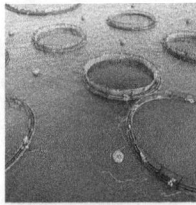	Aquaculture	Nets, PVC pipes, buoys, lines	Rivers, coastline, marine
		Agriculture	Polyhouse, greenhouse sheets, irrigation pipes, fertilizers/pesticide packaging	Rivers, coastline, atmosphere
		Shipping/ Offshore industries		Rivers, marine
		Food and drinks packaging	Containers, plastic bags, bottles, caps, plastics disposable cutlery, etc.	Rivers, coastline
Individual consumers	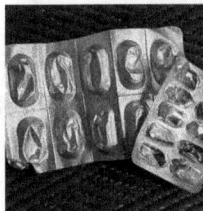	Medicine, personal care products, cosmetics	Microbeads, packaging, toothbrushes, shower loofahs, packaging, etc.	Rivers, coastline
		Clothing and textile	Fibers	Rivers, coastline, atmosphere, marine

Source: Modified from GESAMP (2017).

care and cosmetic products contain microbeads in Macao, China. The use of the same in the region results in the influx of 37 billion microbeads per year into the environment through wastewater treatment plants (Bashir et al. 2021). Plastics sheets that are used for construction of polyhouses, greenhouses, PVC pipes used for irrigation, and fertilizers/pesticide packaging produce microplastics. Plastic mulch is now extensively used in crop production and landscaping to reduce weeds and preserve water, which also results in microplastics. The recovery of plastic mulch is low, and a substantial amount of it remains in the soil after use. Waste from the factories generated during industrial processes, such as surface blasting for cleaning, when discharged into municipal wastewater ends up within natural environments. Fragmentation or abrasion of the plastic products, such as packaging, textiles, and tires, produces tiny plastics debris that passes to the drainage system through rainwater runoff. Further, application of sewage sludge as fertilizer introduces considerable amounts of microplastics into the environment. In fact, roughly every year about 63,000–430,000 and 44,000–300,000 tons of microplastics enter into European and North American croplands, respectively, via sewage sludge emission (Nizzetto et al. 2016b).

1.3   SINK OF MICROPLASTICS: WHERE DO MICROPLASTICS ACCUMULATE?

Microplastics that are released from different emission pathways (Table 1.4) eventually get accumulated into coastal sands, deep sea water and sediments, polar regions, alpine ice sheets, marine species, other secondary consumers, agriculture soil, in addition to other pollutants such as antibiotics, potentially toxic metals, and polycyclic aromatic hydrocarbons. Soils may act as temporary or permanent sinks of microplastics, which postpones or averts the transmission of microplastics to the freshwater system. However, to what extent soil can retain microplastics without affecting the soil quality and productivity is subject of further study. Freshwater ecosystems such as lakes and reservoirs may serve as sinks for microplastic contaminants (Anderson et al. 2016). Moreover, streams, rivers, and estuaries also act as sinks for microplastics, additionally an important transport pathway for terrestrial microplastics into oceans. Benthic sediments are an everlasting sink for microplastics; there is evidence of microplastics deposits in coastal sediments as well as in remote sea basins (Table 1.1). These microplastics particles were detected along the shore of different continents with more abundance in densely populated areas (Browne et al. 2011). There is a resemblance observed between the microplastics found in the marine sediments and sewage effluents with the polyester or acrylic fibers used in the garments worn by the local inhabitants (Browne et al. 2011). Marine environments are now the well-recognized ultimate sink for microplastic particles. However, possible atmospheric transfer of the microplastics cannot be neglected. High concentrations of microplastics have been reported from far-flung uninhabited areas such as the Swiss Alps and the Arctic (Bergmann et al. 2019). Wind systems can possibly transfer these microscopic plastics particles from urban locations over a long range to remote locations.

At some places, low concentrations of the microplastics were observed at the sea surface waters. Adhesion of microplastics to coral reefs has also been revealed (Martin et al. 2019), which indicates a possible sink of microplastics in the coral reefs. Over the last decade, a number of studies reported microplastics deposition in the organs of fishes and other marine organisms (Alfaro-Núñez et al. 2021; Thiele 2021). Since microplastics are very tiny in size and differently colored, both fresh-water and marine organisms, like fishes and zooplanktons, ingest these particles, especially those having small prey size of less than 1 mm. Density of the microplas-tic particles determines availability in the surface water. Biofouling of microplastics affects buoyancy resulting into sinking of the microplastics to the benthos, thus their becoming a "food source" for deposit feeders and detritivores (Wright et al. 2013). Harvesting of the marine organisms for human consumption results in the contami-nation of the food chain with microplastics. Microplastics have been found in fish and fishmeal products meant for human consumption (Thiele et al. 2021). Concentrations of microplastics in marine animals is directly related to the plastics available in the marine water and negatively correlated to the size of microplastics (Zhang 2017).

1.4 HOW ARE SOURCE AND TARGET REGIONS INTERRELATED?

Microplastics of the freshwaters are directly linked to their sources on the terres-trial environment. Wastewater discharged from plastic and allied industries and other sewage effluents are released into rivers as they flow through cities and towns. Microplastic fibers of polyester, polypropylene, acrylic, polyamide, and polyethylene used in garments by the locals accumulate on the shores all over the world. These microfibers largely come from the sewage disposal from the densely populated areas, and habitats are affected with high microplastics contamination (Browne et al. 2011). Terrestrial anthropogenic activities and fisheries activities cause an increase in the abundance of microplastics. For example, mariculture activities were the chief source of the microplastic contamination in Xiangshan Bay in China, with more than 50% of the microplastic fibers derived from mariculture (Chen et al. 2018). Similarly, polyester, polyethylene, and polypropylene fibers were predominant in the Maowei Sea, a popular mariculture bay of China (Zhu et al. 2019). Moreover, the abundance of microplastics in the inflowing rivers to the bay was a little lower than in the Maowei Sea, which suggests that mariculture is also an important contributor to the microplastics apart from the river discharge (Zhu et al. 2019). In the same area, microplastics were found in the gills and gastrointestinal tracts of many fish spe-cies and the soft tissue of oyster samples. Concentrations of microplastics in marine animals is directly related to the plastics available in marine water and negatively correlated to the size of microplastics (Zhang 2017).

1.5 RELATION BETWEEN MICROPLASTICS
PROPERTIES AND TRANSPORT

The majority of plastic waste originates and is disposed of on the land, before reach-ing the marine environment (Figure 1.3). The mass flow of microplastics from land to the marine environment is largely through rivers, but also through air and coastal areas. The primary plastics properties that govern the accumulation and transport are density, surface area, and size of plastics. Further, seawater density, seabed topogra-phy, flow velocity, turbulence, and pressure are the external factors that influence the transport of plastics. Translocation of the plastic particles on the land is determined

FIGURE 1.3 Potential sources and pathways of microplastics.

by the amount of precipitation, hydraulic characteristics, bioturbation, agriculture practices, and the physiochemical properties. Larger microplastics with higher densities as compared to water tend to accumulate in the sediments. A mathematical modeling study of the river Thames hypothetically suggested that a considerable portion (16–38%) of the microplastics are retained on the land, depending on soil sub-catchment features and precipitation patterns. However, microplastics less than 0.2 mm in size are likely to be translocated irrespective of their density (Nizzetto et al. 2016a). The vertical translocation of microplastics into the deep soil and deeper in groundwater is influenced by the size and type of microplastic (Su et al. 2022).

A recent study suggested that microplastics and nanoplastics could translocate vertically into the plants from soil (Luo et al. 2022). Lately, the movement of microplastics from land to the oceans through rivers was monitored theoretically as well as in real field surveys. Lebreton et al.'s (2017) model predicted that every year about 1.15 to 2.41 MMT of plastic debris enters into the ocean through rivers. More extensive field monitoring is required to test the validity of these models and build more such models. Indeed, limited field studies are available at present. Lebreton's estimates rely on data from 13 rivers in seven studies (Lebreton et al. 2017). Meijer et al.'s model used 52 and 84 data points for calibration and validation, respectively, to predict that more than 1000 rivers were responsible for 80% of total annual global emissions (Meijer et al. 2021).

Among the most common microplastic polymers, polyethylene, polypropylene, and ethylene vinyl acetate have low density (<1 g cm^{-3}); polystyrene, polyamide (nylon) have intermediate density (1 to 1.1 g cm^{-3}); and polyacrylonitrile, polyvinyl alcohol, polyester, polyethylene terephthalate, alkyd, polyvinyl chloride, polymethyl methacrylate have high density (>1.1 g cm^{-3}) (Schwarz et al. 2019). There is dynamic interaction of the microplastics at the shoreline, where the mobility of microplastics is determined by physical factors (Zhang 2017).

In the freshwater bodies and oceans there are six major compartments (i.e., beach, epipelagic, and sediment, respectively for both), where microplastics were observed and collected. The concentration of microplastics in any compartment depends significantly on the polymer type. Whereas, change in environmental parameters like salinity and temperature affects the composition within the compartment (Schwarz et al. 2019).

In freshwater, high-density polymers tend to sink at the bottom, whereas low-density polymers float to the surface. In general, plastics with thick walls, low density, and big size are readily transported from rivers to the sea. However, in the rivers and streams, where currents and turbulence are quite high, high-density polymers also become buoyant and are horizontally transported to a large distance, eventually to oceans along the river currents. Additionally, low-density polymers are trapped in the high turbulence, leading to sedimentation. The dams and reservoirs built on rivers also influence the fate and transport of microplastics; due to sedimentation a considerable number of microplastics are retained (Kumar et al. 2021).

The onshore/offshore movement of the microplastics in the marine environment is also influenced by external factors such as source location, geometry, vegetation, tidal regime, and wave direction. Huge quantities of the microplastics remain suspended in the surface and upper mixing layer of sea, where they are transported both horizontally and vertically depending on the hydrodynamic forces. Low temperature and high salinity in the marine environment increase high-density polymers in the water column (Schwarz et al. 2019). Further, the microbial biofilm formation over the microplastics surface (biofouling) and the colonization of plastics polymer by algae or marine invertebrates such as lobsters, crabs, sea urchins, jelly and star fish, increases the density of particles in the freshwater and marine environment, which may affect the translocation. The low-density large microplastics remain suspended at the surface of the oceans until they lose buoyancy through aggregation/biofouling. Microplastics, such as films and fibers, having a larger surface area are more exposed to biofouling, and are therefore likely to have a higher sedimentation rate. Fragmentation and degradation also affect the sedimentation rate of microplastics. The high-density microplastics eventually sink to the bottom of the ocean, subjected to the bottom turbulence, and they may be translocated to the bed sediments or suspended.

1.6   IS MICROPLASTIC A PERSISTENT POLLUTANT?

Over time, the plastic waste just undergoes fragmentation into smaller pieces and will remain present in the environment for hundreds of years. In the marine environment, the half-life of the plastics could range from 58 to 1200 years for high-density polyethylene bottles and pipes respectively (Chamas et al. 2020). Microplastics originate from a diverse range of polymers, and multiple sources may act as the means of transmission for other persistent pollutants. Due to its high sorption capacities, persistent organic pollutants, heavy metals, and pathogenic microbes accumulate over the surface of microplastics. Toxic and persistent chemicals such as pesticides and polychlorinated biphenyls have been reported to interact with and attach to the microplastics, which further increases their impact on the marine ecosystem. The ingestion of such microplastics by marine biota transfers the adsorbed pollutant to the respective organisms and subsequently to the trophic levels. The suspended microplastics are more persistent as they remain present for longer and readily available for the marine organisms. As mentioned earlier, the buoyancy of microplastics also determines if the microplastics will remain suspended or settle down.

1.7   GLOBAL TREND

Until September 2019, there were a total of 2501 publication records on microplastics in the Web of Science Core Collection (WoSCC) database (Zhang et al. 2020). According to an estimate, every year between 0.013 and 25 MMT microplastics and nanoplastics enter into the marine atmosphere and are released into the oceans (Allen et al. 2022). Most workers identify plastics of the size ranging from 1 μm to 5 mm as microplastics irrespective of the primary or secondary source of origin. In the marine environment plastics are the major (60–80%) solid debris (Yaranal et al. 2021). Polyvinyl chloride, polyethylene, polystyrene, polypropylene, nylon, polyethylene terephthalate, polyamide, and acrylonitrile butadiene styrene are some of the commonly found microplastics. The number of studies on microplastics has increased significantly since 2011 (Zhang et al. 2020). First recorded in the marine environment, microplastics research on marine systems and marine organisms is still dominant. Microplastics are investigated more extensively in the marine environment than in other ecosystems (Zhang et al. 2020). Some researchers studied the presence and translocation of microplastics on land (Nizzetto et al. 2016a; Su et al. 2022; Luo et al. 2022). The microplastics issues extend beyond the oceans. In order to make a better global presentation and to propose strategies and control policies, full information on the sources, pathways, and sinks of microplastics should be prepared and available. There is a need to broaden microplastics research beyond the marine to the terrestrial environment. Recently, microplastics were detected in human blood. The exact impact of microplastics on human health should be evaluated. Future studies should include exposure pathways, potential ecotoxicological impact on human health and marine life, and the safety threshold of microplastics should be determined (Zhang et al. 2020).

1.8   SUMMARY

Plastics are an indispensable part of our daily lives, ubiquitously present everywhere. By 2050, there will be 12,000 MMT tons of plastic garbage in the landfills and the environment. Microplastics, the tiny or microscopic particles of plastic smaller than 5 mm, are the most frequently detected plastics worldwide that have been identified as a widespread global threat to human health and the environment. In the marine environment, microplastics have been detected globally from different and far-reaching locations such as Arctic regions, different coastal areas, river estuaries, etc. For any country, the size of the population and the percentage of waste managed can tell us a lot about the plastics waste generated by the country and its plastics contribution to the marine environment. Every year, between 0.013 and 25 MMT microplastics are potentially transmitted into the marine atmosphere, and released into the oceans. Oceans are the ultimate sink for microplastics. Once released into the environment, microplastics become an inseparable part of marine ecosystems. Microplastics can withstand varying environmental conditions, resist environmental degradation, and will last long in the environment after discharge.

Microplastics can be divided into primary and secondary microplastics. The breakdown of larger plastic by means of chemical, mechanical, and biological degradation in the environment generates secondary microplastics, which are largely composed of parent polymers. Primary microplastics include cosmetics, personal care products, drugs, cleansers, paints, gels, etc. They are usually manufactured as microbeads and are purposely added to consumer products for application.

Worldwide, only a fraction (1%) of the plastic waste enters directly into the marine environment. Most of the emissions are to terrestrial land and freshwaters either as primary or secondary. Some of the identified sources of marine microplastics pollution include pellets and fragments used in the plastic manufacturing, fabricating, and recycling industries; improper disposal of solid waste, sewage, and wastewater; plastics used in the construction industries; pellets, tire dust, and road paint abrasions generated by vehicles; consumer goods, packaging, and microbeads from the tourism industries. Containers, plastic bags, bottles, caps, plastic disposable cutlery, etc. used for food and drinks packaging; textile fibers from individual consumers and the textile industries; sports, fisheries, aquaculture, agriculture, shipping/offshore industries are also the sources of microplastics. Atmospheric microplastics are an important part of the marine plastic cycle.

Fragmentation or abrasion of plastic products, such as from packaging, textiles, and tires, produces tiny plastics debris that passes to the drainage system through rainwater runoff. Microplastics that are released from different emission pathways are subsequently translocated and accumulated into coastal sands, deep sea water and sediments, polar regions, alpine ice sheets, marine species, other secondary consumers, and agriculture soil. Since microplastics are very tiny in size and differently colored, both freshwater and marine organisms, like fishes and zooplanktons, ingest these particles, especially those having small prey size of less than 1 mm. The microplastics consumed by marine organisms subsequently transfer to the next trophic level and so on.

Microplastics in freshwater are directly linked to their sources from the terrestrial environment. The mass flow of microplastics from land to the marine environment is largely through rivers, but also through air and coastal areas. The primary plastic properties that govern accumulation and transport are density, surface area, and size of plastics. In the marine environment, mobility of microplastics is influenced by physical factors such as source location, geometry, vegetation, tidal regime, and wave direction. Biofouling also affects the translocation.

Microplastics originate from a diverse range of polymers and multiple sources, and may act as a means of transmission for other persistent pollutants. Due to their high sorption capacities, persistent organic pollutants, heavy metals, and pathogenic microbes accumulate over the surface of microplastics. The number of studies on microplastics has increased significantly since 2011. Microplastics pollution is more extensively studied in the marine environment than in other ecosystems. Sources and sinks, pathways regarding microplastics are important in order to prepare global prevention and control strategies.

REFERENCES

Alfaro-Núñez, A., Astorga, D., Cáceres-Farías, L., Bastidas, L., Soto Villegas, C., Macay, K. and Christensen, J. H. 2021. Microplastic pollution in seawater and marine organisms across the Tropical Eastern Pacific and Galápagos. *Scientific Reports*, 11(1):1–8.

Allen, D., Allen, S., Abbasi, S., et al. 2022. Microplastics and nanoplastics in the marine-atmosphere environment. *Nature Reviews Earth and Environment*. doi: 10.1038/s43017-022-00292-x.

Anderson, J. C., Park, B. J. and Palace, V. P. 2016. Microplastics in aquatic environments: Implications for Canadian ecosystems. *Environmental Pollution*, 218:269–280.

Arthur, C., Baker, J. E. and Bamford, H. A. 2009. The occurrence, effects, and fate of microplastic marine debris. In: *Proceedings of the International Research Workshop*, 9–11 September, 2008. University of Washington Tacoma, Tacoma, WA, USA.

Bashir, S. M., Kimiko, S., Mak, C. W., Fang, J. K. H. and Gonçalves, D. 2021. Personal care and cosmetic products as a potential source of environmental contamination by microplastics in a densely populated Asian city. *Frontiers in Marine Science*, 8:604.

Bergmann, M., Mützel, S., Primpke, S., Tekman, M. B., Trachsel, J. and Gerdts, G. 2019. White and wonderful? Microplastics prevail in snow from the Alps to the Arctic. *Science Advances*, 5(8):1157.

Blott, S. J. and Pye, K. 2001. GRADISTAT: A grain size distribution and statistics package for the analysis of unconsolidated sediments. *Earth Surface Processes and Landforms*, 26(11):1237–1248.

Browne, M. A., Crump, P., Niven, S. J., Teuten, E., Tonkin, A., Galloway, T. and Thompson, R. 2011. Accumulation of microplastic on shorelines worldwide: Sources and sinks. *Environmental Science and Technology*, 45(21):9175–9179.

Chae, D. H., Kim, I. S., Kim, S. K., Song, Y. K. and Shim, W. J. 2015. Abundance and distribution characteristics of microplastics in surface seawaters of the Incheon/Kyeonggi coastal region. *Archives of Environmental Contamination and Toxicology*, 69(3):269–278.

Chamas, A., Moon, H., Zheng, J., Qiu, Y., Tabassum, T., Jang, J. H., Abu-Omar, M., Scott, S. L. and Suh, S. 2020. Degradation rates of plastics in the environment. *ACS Sustainable Chemistry and Engineering*, 8(9):3494–3511.

Chen, M., Jin, M., Tao, P., Wang, Z., Xie, W., Yu, X. and Wang, K. 2018. Assessment of microplastics derived from mariculture in Xiangshan Bay, China. *Environmental Pollution*, 242:1146–1156.

Choudhary, S., Neelavanan, K. and Saalim, S. M. 2022. Microplastics in the surface sediments of Krossfjord-Kongsfjord system, Svalbard, Arctic. *Marine Pollution Bulletin*, 176:113452.

Claessens, M., De Meester, S., Van Landuyt, L., De Clerck, K. and Janssen, C. R. 2011. Occurrence and distribution of microplastics in marine sediments along the Belgian coast. *Marine Pollution Bulletin*, 62(10):2199–2204.

Coffin, S. 2020. Proposed definition of 'microplastics in drinking water'. *California Water Boards*. www.waterboards.ca.gov/drinking_water/certlic/drinkingwater/docs/stffrprt_jun3.pdf

Cole, M., Lindeque, P., Halsband, C. and Galloway, T. S. 2011. Microplastics as contaminants in the marine environment: A review. *Marine Pollution Bulletin*, 62(12):2588–2597.

De-la-Torre, G. E., Dioses-Salinas, D. C., Castro, J. M., Antay, R., Fernández, N. Y., Espinoza-Morriberón, D. and Saldaña-Serrano, M. 2020. Abundance and distribution of microplastics on sandy beaches of Lima, Peru. *Marine Pollution Bulletin*, 151:110877.

EFSA Panel on Contaminants in the Food Chain (CONTAM). 2016. Presence of microplastics and nanoplastics in food, with particular focus on seafood. *EFSA Journal*, 14(6):e04501.

Ershova, A., Makeeva, I., Malgina, E., Sobolev, N. and Smolokurov, A. 2021. Combining citizen and conventional science for microplastics monitoring in the White Sea basin (Russian Arctic). *Marine Pollution Bulletin*, 173:112955.

Frias, J. P. and Nash, R. 2019. Microplastics: Finding a consensus on the definition. *Marine Pollution Bulletin*, 138:145–147.

Galgani, F., Hanke, G., Werner, S. D. and De Vrees, L. 2013. Marine litter within the European marine strategy framework directive. *ICES Journal of Marine Science*, 70(6):1055–1064.

Gao, F., Li, J., Hu, J., Sui, B., Wang, C., Sun, C., Li, X. and Ju, P. 2021. The seasonal distribution characteristics of microplastics on bathing beaches along the coast of Qingdao, China. *Science of the Total Environment*, 783:146969.

GESAMP. 2015. Sources, fate and effects of microplastics in the marine environment (part 1). www.gesamp.org/publications/reports-and-studies-no-90.

GESAMP. 2017. Sources, fate and effects of microplastics in the marine environment: Part two of a global assessment. www.gesamp.org/site/assets/files/1275/sources-fate-and-effects-of-microplastics-in-the-marine-environment-part-2-of-a-global-assessment-en.pdf.

Geyer, R., Jambeck, J. R. and Law, K. L. 2017. Production, use, and fate of all plastics ever made. *Science Advances*, 3(7):e1700782.

Hartmann, N. B., Huffer, T., Thompson, R. C., Hassellöv, M., Verschoor, A., Daugaard, A. E., Rist, S., Karlsson, T., Brennholt, N., Cole, M. and Herrling, M. P. 2019. Are we speaking the same language? Recommendations for a definition and categorization framework for plastic debris. *Environmental Science and Technology*, 53(3):1039–1047.

Hernandez, E., Nowack, B. and Mitrano, D. M. 2017. Polyester textiles as a source of microplastics from households: A mechanistic study to understand microfiber release during washing. *Environmental Science and Technology*, 51(12):7036–7046.

Jambeck, J. R., Geyer, R., Wilcox, C., Siegler, T. R., Perryman, M., Andrady, A., Narayan, R. and Law, K. L. 2015. Plastic waste inputs from land into the ocean. *Science*, 347(6223):768–771.

Jorquera, A., Castillo, C., Murillo, V., Araya, J., Pinochet, J., Narváez, D., Pantoja-Gutiérrez, S. and Urbina, M. A. 2022. Physical and anthropogenic drivers shaping the spatial distribution of microplastics in the marine sediments of Chilean fjords. *Science of the Total Environment*, 814:152506.

Kaviarasan, T., Dhineka, K., Sambandam, M., Sivadas, S. K., Sivyer, D., Hoehn, D., Pradhan, U., Mishra, P. and Murthy, M. R. 2022. Impact of multiple beach activities on litter and microplastic composition, distribution, and characterization along the southeast coast of India. *Ocean and Coastal Management*, 223:106177.

Kedzierski, M., Le Tilly, V., Bourseau, P., Bellegou, H., César, G., Sire, O. and Bruzaud, S. 2016. Microplastics elutriation from sandy sediments: A granulometric approach. *Marine Pollution Bulletin*, 107(1):315–323.

Kumar, R., Sharma, P., Verma, A., Jha, P. K., Singh, P., Gupta, P. K., Chandra, R. and Prasad, P. V. 2021. Effect of physical characteristics and hydrodynamic conditions on transport and deposition of microplastics in riverine ecosystem. *Water*, 13(19):2710.

La Daana, K. K., Officer, R., Lyashevska, O., Thompson, R. C. and O'Connor, I. 2017. Microplastic abundance, distribution and composition along a latitudinal gradient in the Atlantic Ocean. *Marine Pollution Bulletin*, 115(1–2):307–314.

Lebreton, L., Van Der Zwet, J., Damsteeg, J. W., Slat, B., Andrady, A. and Reisser, J. 2017. River plastic emissions to the world's oceans. *Nature Communications*, 8(1):1–10.

Luo, Y., Li, L., Feng, Y., Li, R., Yang, J., Peijnenburg, W. J. and Tu, C. 2022. Quantitative tracing of uptake and transport of submicrometre plastics in crop plants using lanthanide chelates as a dual-functional tracer. *Nature Nanotechnology*, 1–8.

Martin, C., Corona, E., Mahadik, G. A. and Duarte, C. M. 2019. Adhesion to coral surface as a potential sink for marine microplastics. *Environmental Pollution*, 255:113281.

Meijer, L. J., van Emmerik, T., van der Ent, R., Schmidt, C. and Lebreton, L. 2021. More than 1000 rivers account for 80% of global riverine plastic emissions into the ocean. *Science Advances*, 7(18):eaaz5803.

Menéndez-Pedriza, A. and Jaumot, J. 2020. Interaction of environmental pollutants with microplastics: A critical review of sorption factors, bioaccumulation and ecotoxicological effects. *Toxics*, 8(2):40.

Mistri, M., Scoponi, M., Granata, T., Moruzzi, L., Massara, F. and Munari, C. 2020. Types, occurrence and distribution of microplastics in sediments from the northern Tyrrhenian Sea. *Marine Pollution Bulletin*, 153:111016.

Ng, K. L. and Obbard, J. P. 2006. Prevalence of microplastics in Singapore's coastal marine environment. *Marine Pollution Bulletin*, 52(7):761–767.

Nizzetto, L., Bussi, G., Futter, M. N., Butterfield, D. and Whitehead, P. G. 2016a. A theoretical assessment of microplastic transport in river catchments and their retention by soils and river sediments. *Environmental Science: Processes and Impacts*, 18(8):1050–1059.

Nizzetto, L., Futter, M. and Langaas, S. 2016b. Are agricultural soils dumps for microplastics of urban origin? *Environmental Science and Technology*, 50(20):10777–10779.

Renzi, M., Blašković, A., Broccoli, A., Bernardi, G., Grazioli, E. and Russo, G. 2020. Chemical composition of microplastic in sediments and protected detritivores from different marine habitats (Salina Island). *Marine Pollution Bulletin*, 152:110918.

Saeed, T., Al-Jandal, N., Al-Mutairi, A. and Taqi, H. 2020. Microplastics in Kuwait marine environment: Results of first survey. *Marine Pollution Bulletin*, 152:110880.

Schwarz, A. E., Ligthart, T. N., Boukris, E. and Van Harmelen, T. 2019. Sources, transport, and accumulation of different types of plastic litter in aquatic environments: A review study. *Marine Pollution Bulletin*, 143:92–100.

Sobhani, Z., Lei, Y., Tang, Y., Wu, L., Zhang, X., Naidu, R., Megharaj, M. and Fang, C. 2020. Microplastics generated when opening plastic packaging. *Scientific Reports*, 10(1):1–7.

Su, L., Xiong, X., Zhang, Y., Wu, C., Xu, X., Sun, C. and Shi, H. 2022. Global transportation of plastics and microplastics: A critical review of pathways and influences. *Science of the Total Environment*, 154884.

Suzuki, G., Uchida, N., Tanaka, K., Matsukami, H., Kunisue, T., Takahashi, S., Viet, P. H., Kuramochi, H. and Osako, M. 2022. Mechanical recycling of plastic waste as a point source of microplastic pollution. *Environmental Pollution*, 303:119114.

Thiele, C. J., Hudson, M. D., Russell, A. E., Saluveer, M. and Sidaoui-Haddad, G. 2021. Microplastics in fish and fishmeal: An emerging environmental challenge?. *Scientific Reports*, 11(1):1–12.

Thompson, R. C., Olsen, Y., Mitchell, R. P., Davis, A., Rowland, S. J., John, A. W., McGonigle, D. and Russell, A. E. 2004. Lost at sea: Where is all the plastic?. *Science*, 304(5672):838–838.

Van Cauwenberghe, L., Devriese, L., Galgani, F., Robbens, J. and Janssen, C. R. 2015. Microplastics in sediments: A review of techniques, occurrence and effects. *Marine Environmental Research*, 111:5–17.

Venkatramanan, S., Chung, S. Y., Selvam, S., Sivakumar, K., Soundhariya, G. R., Elzain, H. E. and Bhuyan, M. S. 2022. Characteristics of microplastics in the beach sediments of Marina tourist beach, Chennai, India. *Marine Pollution Bulletin*, 176:113409.

Wright, S. L., Thompson, R. C. and Galloway, T. S. 2013. The physical impacts of microplastics on marine organisms: A review. *Environmental Pollution*, 178:483–492.

Yaranal, N. A., Subbiah, S. and Mohanty, K. 2021. Identification, extraction of microplastics from edible salts and its removal from contaminated seawater. *Environmental Technology and Innovation*, 21:101253.

Zaki, M. R. M., Ying, P. X., Zainuddin, A. H., Razak, M. R. and Aris, A. Z. 2021. Occurrence, abundance, and distribution of microplastics pollution: An evidence in surface tropical water of Klang River estuary, Malaysia. *Environmental Geochemistry and Health*, 43(9):3733–3748.

Zhang, H. 2017. Transport of microplastics in coastal seas. *Estuarine, Coastal and Shelf Science*, 199:74–86.

Zhang, W., Zhang, S., Wang, J., Wang, Y., Mu, J., Wang, P., Lin, X. and Ma, D. 2017. Microplastic pollution in the surface waters of the Bohai Sea, China. *Environmental Pollution*, 231:541–548.

Zhang, Y., Pu, S., Lv, X., Gao, Y. and Ge, L. 2020. Global trends and prospects in microplastics research: A bibliometric analysis. *Journal of Hazardous Materials*, 400:123110.

Zhao, S., Zhu, L. and Li, D. 2015. Microplastic in three urban estuaries, China. *Environmental Pollution*, 206:597–604.

Zhu, J., Zhang, Q., Li, Y., Tan, S., Kang, Z., Yu, X., Lan, W., Cai, L., Wang, J. and Shi, H. 2019. Microplastic pollution in the Maowei Sea, a typical mariculture bay of China. *Science of the Total Environment*, 658:62–68.

Ziajahromi, S., Neale, P. A., Rintoul, L. and Leusch, F. D. 2017. Wastewater treatment plants as a pathway for microplastics: Development of a new approach to sample wastewater-based microplastics. *Water Research*, 112:93–99.

2 Analytical Methods for the Identification and Assessment of Microplastics

Marine plastics and microplastics pollution is gradually affecting the marine environment due to the large number of plastics produced in many sectors such as packaging, construction, and transportation, and their eventual transmission and accumulation into the oceans. For better understanding and limiting the effects of microplastics, monitoring and evaluation of their occurrence and distribution across various environments, including marine, are essentially required. This is also important for the policy makers to design appropriate mitigation and preventive policies. Standardized processes and approaches should be used to establish uniformity and comparable data on microplastics occurrence across varied samples and locations in studies all over the world. This chapter identifies and provides an overview of the methods available for measuring microplastics in the marine environment. To ensure global comparability of microplastics monitoring, a high level of quality assurance is required at each step including sampling, separation, preprocessing, and identification.

2.1 SAMPLING

The collection of samples from the site of occurrence is the first and foremost step to studying microplastics distribution and occurrence. The marine environment can be divided into sub-components, including sea surface, water column, sediment, and marine organisms. The sampling can be bulk, selective, or volume-reduced depending on the target sub-component and research objective (Hidalgo-Ruz et al. 2012; Liu et al. 2020).

2.1.1 SELECTIVE SAMPLING

Selective sampling is the direct sampling of the particles visible to the human eye from the environment. The method is frequently used for the sampling of plastics particles on the surface of sediments, which are mostly in the size range of 1–6 mm and are easily recognizable. However, microplastics often remain combined with other forms of debris to become indistinguishable. Therefore, special caution and attention should be exercised in the field while sampling, to reduce the chance of overlooking.

DOI: 10.1201/9781003312086-2

2.1.2 BULK SAMPLING

Bulk sampling is a large sampling where the complete volume of the sample is acquired without any size reduction. This type of sampling is preferred when microplastics cannot be recognized visually since they are covered by sediment, quality is very low, or their size is too tiny to be identified visibly with the eyes.

2.1.3 VOLUME-REDUCED SAMPLING

In this type of method, the volume of sample is reduced by concentrating the samples at the site itself. Water is usually filtered through nets and the sediment samples are sieved on the beach itself or on the ship, and the retained microplastic particles are further processed and identified in the lab.

Since microplastics are comparatively low in concentrations in the marine environment, sampling often becomes more challenging for precise estimation and identification. Normally, a huge volume of samples is required for better monitoring. For the marine surface water and suspended microplastic particles, bongo/zooplankton nets, plankton samplers, neuston nets, and manta trawls are commonly used, while for the shallow bottom water and sediments, bottom trawls can be used (Table 2.1). The mesh size may vary depending on the target particle size. A flowmeter is placed at the mouth of the net to record the volume water filtered through the net. This is important for the data normalization and further calculating microplastic concentrations (items/grams) per unit volume of water. The selection of the sampling equipment may also vary with the sea conditions (Beaufort scale); for example, a manta trawl can be used in calm water while neuston catamarans can work in high waves, escaping wave hopping and equipment damage. Deep-sea sampling requires more advanced and supplicated vessels (e.g., with special attachments or grabs multicorers; a ROV Victor 6000 can be used having a reach of up to 5000 m) (van Cauwenberghe et al. 2013).

Marine organisms were found to ingest microplastics; therefore, it is important to monitor microplastic uptake in marine organisms and assess its impact. The sampling technique of the marine organisms may vary with the target organism, the ecological habitat, and the study objective. For instance, the benthic invertebrates can be collected in grabs, creels, and traps, or through bottom trawl. The nektonic, planktonic and invertebrates can be collected using manta and bongo nets. The marine fishes can be trapped in benthic trawls from the surface and midwater (Lusher et al. 2017). If the organism is too big, such as a shark, whale, big fishes, the digestive tract can be removed and stored, while small organisms such as mussels and barnacles can be evaluated completely for the presence of microplastics (Mai et al. 2018).

The beach side sediments are considerably easy to collect with the help of tweezers, tablespoons, small shovels, trowels, or hands (Löder and Gerdts 2015; Malankowska et al. 2021). A quadrat frame or a corer can be used to demarcate an area of sampling and a non-plastic container or bag to collect and store the samples. Following sampling, the preliminary screening can be done on the site itself, and the screened microplastics can be brought to the laboratory for further examination. The abundance is usually normalized according to the sampling area, weight, or volume

TABLE 2.1

Common instruments used for the sampling of the microplastics in the marine environment.

	Instrument		Type of samples
1	Bongo/zooplankton net		Floating and suspended particles
2	Plankton samplers		Floating particles
3	Neuston net		First few centimeters of the water column
4	Bottom trawl		Bottom water and sediments
5	Manta trawl		Surface of ocean

(Continued)

TABLE 2.1
(Continued)

	Instrument		Type of samples
6	ROV Victor 6000		Deep sea water and sediments
7	Multicorer		Deep sea water and sediments
8	Ekman grab		Sediments

of the sediments, and units may vary as per the method of sampling followed. The sampling depth is usually the top 5 cm while a higher sampling depth is also reported in the literature.

2.2   SEPARATION OF MICROPLASTICS FROM SAMPLES

The microplastics must be separated from the samples after collection in order to get clean particles for subsequent quantification and identification. Filtration or sieving, density separation, and digestive processes are the usual separation procedures (Table 2.2).

2.2.1   FILTRATION OR SIEVING

The water and sediment samples are typically passed through sieves and filters of different mess size to separate microplastic particles from the samples. The tools are usually made of stainless-steel screens or fiberglass filters (Lv et al. 2021). In the sieving method, water and sediment samples are passed through sieves of different mesh sizes (0.038–4.75 mm) to remove larger particles and other impurities

TABLE 2.2

Advantages and limitations of the common methods used in microplastics extraction from marine sediments.

Operation	Reagent or material	Advantages	Drawbacks
Separation	NaCl saturated	Availability and low price; largest used, data comparable; green method to the environment	Loss of MP possible due to lowest density
	$ZnCl_2$	High recovery	Expensive; hazardous to operator and environment
	NaI	High recovery; reuse possible	Expensive; sensitive to pH
Isolation	Hand sorting	Simple and easy to work	Time consuming, experiment required; limited on MP large size
	Sieving	Rapid; easy to clean and reuse	MP size range still limited; loss of small MP; manually required later
	Filtration	Highly MP retained; diverse filters are available; largest used, data comparable	Depending strongly to digestion and separation step; filter type must be defined
Digestion	H_2O_2	Inexpensive, availability; largest used, data comparable	Instability; loss of MP color observed; boil intensely at high temperature; adding slowly due to gas bubbles during heated
	H_2O_2 and Fe^{2+}	Fast and efficient; MP characteristics preserved	Sharing drawbacks with H_2O_2 solution; precipitation possible
	Enzyme	Highly efficient; no damage to sensitive polymers	Time consuming; complexity, pH required

Source: Phuong et al. (2021).

(Hidalgo-Ruz et al. 2012; Lv et al. 2021). The particles that pass through the sieve are either discarded or passed again through a smaller mesh size, while the remaining particles are collected and sorted subsequently. Sequential passing samples through different mesh sizes allow particle separation particles into multiple size groups. The size of microplastics is determined by the sieve mesh size or the filter pore size. Silt samples can be passed through a larger sieve prior to downstream sieving. Sieving can be wet or dry depending on the nature of the sample. If most of the particles are less than 500 μm in size, the sediment samples can be dried before sieving to reduce the subsequent extraction volume (Löder and Gerdts 2015). Further, handling of the sample during the process of sieving should be done with utmost care so that additional particles may not form by the mechanical fragmentation from larger brittle plastic material. The number of sieves and mesh size may change according to the sieving goal, like selection of a specific size range of microplastics.

The sieving is usually followed by filtering; however filtering can be excluded if secondary removal is not required (Hidalgo-Ruz et al. 2012). In general, the pore size of the filters may vary between 1–2 μm (Hidalgo-Ruz et al. 2012; Lv et al. 2021).

Filtration unit may be aided with a vacuum to facilitate microplastics extraction from a large volume of water samples in less time. A small portion of the microplastics may adhere to the wall of the filter unit, resulting in sample loss. To minimize loss, the container walls can be washed onto the filter. For accurate estimates, it is also important to avoid external contamination when taking samples and processing them in the lab. The application of filtration and sieving for microplastics separation has some limitations, for instance, clogging of pore or mesh due to the minerals and organic particles present in the samples, especially when the pore or mesh is very small. On the other hand, smaller size particles are lost when the pore or mesh size is larger. This may lead to underestimation of the abundance of microplastics (Elizalde-Velázquez and Gómez-Oliván 2021). Thus, there is a requirement to standardize the pore or mesh size in the protocols to precisely separate microplastics.

2.2.2 DENSITY SEPARATION

Sediment samples cannot be directly filtered due to the presence of more sediment particles as compared to the microplastics. Therefore, density-based separation is the frequently used technique for the separation of microplastic particles from marine sediments by floatation, followed by filtration of the supernatant (Pagter et al. 2018).

The densities of common microplastic polymers are typically less than the sediments. Therefore, microplastics can be separated by flotation with saturated salt solutions of considerably high density (Löder and Gerdts 2015). For instance, density of silicone is approximately 0.8 g cm^{-3}, and polyethylene terephthalate (PET) is 1.35–1.40 g cm^{-3} and polyvinyl chloride (PVC) is 1.30–1.48 g cm^{-3}; further expanded plastic foams have much lesser densities than polymers; expanded polystyrene (EPS) has a density of less than 0.05 g cm^{-3}. However, sediments have much higher densities of 2.65 g cm^{-3} (Pagter et al. 2018).

The most commonly used liquids for the density-based separations are NaCl, ZnCl$_2$, and NaI solutions with a density range of 1.15–1.3 g cm^{-3}, 1.5–1.8 g cm^{-3}, and 1.55–1.8 g cm^{-3}, respectively (Cutroneo et al. 2021). Other salts and solution densities reported in various studies are CaCl$_2$ (1.3–1.35 g cm^{-3}), ZnBr$_2$ (1.7 g cm^{-3}), NaBr (1.37 g cm^{-3}), 3Na$_2$WO$_4$·9WO$_3$·H$_2$O (1.4–1.65 g cm^{-3}), Na$_2$WO$_4$·2H$_2$O (1.4 g cm^{-3}), Li$_2$WO$_4$ (1.62 g cm^{-3}), KI (1.7 g cm^{-3}), and NaH$_2$PO$_4$ (1.4–1.45 g cm^{-3}) (Cutroneo et al. 2021). However, most of the salts used for density-based separation are not good for humans or the environment. Further, some of the salts such as sodium polytungstate are quite expensive. A suitable salt solution for density separation must have high density, be low in cost, and be no hazard to the environment. To separate microplastic particles from less dense biological materials, 96% ethanol can be used as a substitute of the salt solution with better results than H$_2$O$_2$ digestion (Herrera et al. 2018). However, it has to be noted that solvents like ethanol may dissolve or harm certain kinds of plastics, mostly microscopic particles.

For millimeter-size microplastics, density-based flotation may provide good separation of particles. However, for the particles that are much smaller in size, flotation is not suitable since buoyant-force is quite low for smaller particles, and surface-fouling may also affect particle density significantly (Nguyen et al. 2019). Another modified approach of density-based separation is the Munich plastic sediment separator. It uses a high-density separation fluid which allows separation of large plastic particles (5–20 mm mesoplastics and 1–5 mm large microplastics) as well as small microplastics less than 1 mm in size (Imhof et al. 2012). After density-based separation, the micro- and macroparticles are usually passed through a filter with pore sizes from 1 to 2 µm, using a vacuum for further size-based separation (Malankowska et al. 2021).

2.2.3 OTHER METHODS

Electrostatic separators called hamos separators can be used to sort plastics and other materials as well as metals (Felsing et al. 2018). This approach does not have the limitation of particle densities, but instead uses electrostatic properties in different recycling or separating processes. The conductive sediments and other particulate matter can be separated from non-conductive microplastics due to the difference in conductive properties. The samples are properly dried prior to separation and placed on a vibrating conveyor to a rotating grounded metal drum and transported to a electrostatically charged electrode with up to 35 kV (Felsing et al. 2018). Another method used to separate microplastics from samples is based on the principle of hydrophobicity (Nguyen et al. 2019). In froth flotation, hydrophobic particles stick to the bubble surface that moves particles to the air-liquid interface. However, as compared to mineral separation, for plastic separation froth flotation may not be suitable as inconsistence in bubbles can result in high particle losses (up to 45% in a study) (Imhof et al. 2012; Nguyen et al. 2019). Microparticles of sizes 1–10 µm can be detected using the particle-electrode impact method that comprises a three-electrode system that checks the electrochemical reaction—the collisions of the particles with the electrodes caused by Brownian motion (Shimizu et al. 2017). Flow cytometry coupled with a visual stochastic network embedding (viSNE) can facilitate detection of small microplastics (1–20 µm) in the water samples (Sgier et al. 2016). Moreover, there are certain limitations of using these methods, such as requirement of specialized instrument, too much time per sample, no information on the type of polymer, saturation of device, and fluctuations in surface charge due to biofouling and weathering of plastic (Prata et al. 2019). Another technique for particle sorting is Pinched Flow Fractionation, which uses two separate inlets, one from where fluid-containing particles go inside, and from where fluid is pinched through at a higher flow rate. The particles of different sizes flow through different streamlines based on their center of mass leading to separation, and particles are trapped in dedicated reservoirs (Elsayed et al. 2021).

2.3 PREPROCESSING

The samples collected from the marine environment may contain lot of impurities such as biological materials, microbial biofilms, detritus, and algae, which can interfere in the subsequent analysis. Therefore, it is essential to separate and extract the samples. As discussed before, the samples can be primarily sorted by filtration, visual screening, and density separation. Further, adhered biological materials require special preprocessing, mainly by digestion using acid, alkali, oxidizing agents, or enzymes (Prata et al. 2019; Gong and Xie 2020).

2.3.1 REMOVAL OF ORGANIC AND BIOLOGICAL MATTER

The marine sediment samples may contain organic contaminant up to 0.5–7.0% that may influence further analysis (Prata et al. 2019). Biotic matter in the samples can be mistaken as microplastics that may lead to overestimation of microplastics. Therefore, the digestion method facilitates breakdown of biotic materials without changing plastic particles' chemical or structural integrity (Prata et al. 2019). However, digestion is not always necessary; for instance, when there is low quantity of organic matter in the samples. The recovery rate should also be taken into account, considering the digestion method for the study. Different microplastics have varied chemical resistance, thus there could be treatment effects of the acid or alkali used for digestion. Common digestion methods for the removal of organic matter, their efficacy and recovery rates are presented in Table 2.3 (Prata et al. 2019).

The microplastics ingested by the marine biota can be isolated and identified after dissection, homogenization, depuration, and digestion of the organism tissues with chemicals or enzymes (Lusher et al. 2017). Dissection of the internal organs is a traditional and cheaper method for the visual identification of particles larger than 0.5 mm in the gastrointestinal tracts and other organs, but much smaller particles can translocate to tissues that require additional processing such as fixation and cryosection (Nguyen et al. 2019). Commonly, microplastics are investigated in the stomach or the gastrointestinal tract of the bigger organisms while smaller organisms can be investigated entirely (Miller et al. 2017). Further, in the laboratory it is relatively easy to study the impact of some particular microplastics on the specific species; however in most of the field studies predominantly the digestive tract or stomach is investigated (Lusher et al. 2017; Stock et al. 2019). Detection of microplastics in the marine organism is a lengthy process, therefore requiring preservation of the organism until further analysis (Löder and Gerdts 2015; Nguyen et al. 2019). Biota samples should be washed to avoid external plastics contamination; for living biota depuration should be done for cleaning the gut; the samples that cannot be processed immediately must be frozen or preserved in suitable fixative (Stock et al. 2019). Until processing, the samples can be kept at −20°C or can be preserved in formaldehyde-based fixatives such as 10% formalin, 37% formaldehyde, or 4% Baker's calcium formol; otherwise, in 70–80% ethanol or 99% bidistilled glycerin (Miller et al. 2017). There is no universal guideline available for the processing of biota samples; dissection or depuration steps may vary with the nature of the sample. Nonetheless, representative indicator species should be investigated from the sampling locations, and biomarkers and pollutants concentrations should be assessed for interpreting the results properly.

One option to remove organic materials is acid digestion. However, some polymers have limited acid resistance and could degrade, especially at high temperature and concentrations (e.g., polyamide and polyethylene terephthalate) (Table 2.3). The destruction of organic matter could be up to 100% with a recovery rate of <95%. Most of the polymers, including PE, HDPE, PS, polyester, and PVC, remain unaffected by the prolonged (1 month) treatment of 55% HNO_3 at room temperature; only nylon degraded and PVC whitened (Naidoo et al. 2017). HCl is not recommended since it does not destroy all organic matter and is inefficient in digestion (Cole et al. 2014). Acid digestion should be used with caution as it can result in the underestimation of microplastics (Prata et al. 2019). Another option of digestion is the use of alkali such as NaOH or KOH in different concentrations and temperature at variable duration (Table 2.3). Increasing the molarity and the temperature could enhance digestion; however polymers such as polycarbonates, cellulose acetate, polyethylene terephthalate, and polyvinyl chloride could be degraded (Stock et al. 2019). KOH is recommended as a suitable solvent to study microplastics ingested by marine biota; it readily dissolves most organic material without affecting most plastic types except cellulose acetate (Kühn et al. 2017).

Hydrogen peroxide (H_2O_2) is an efficient and widely used oxidizing agent for removing organic matter, even better than NaOH and HCl (Nuelle et al. 2014). H_2O_2 (30–35%) has very little to no effect on the polymers including PVC, PET, nylon, ABS, PC, PUR, PP, LDPE, LLDPE (linear LDPE), HPDE; further, nylon can degrade and the color of PET can change with H_2O_2 (35%) treatment at 50°C for 96 h (Nuelle et al. 2014; Karami et al. 2017). Sodium hypochlorite (NaClO) is inefficient (digestibility <95%) in digestion of biological matter, thus could not be used as a oxidizing agent at temperatures ranging from 25°C to 60°C (Karami et al. 2017). Digestion may produce foam that could affect low recovery of microplastics. Moreover, incubation temperature can also affect efficiency of H_2O_2 (Prata et al. 2019). Enzymatic digestion is considered as an alternative to chemical digestion and comparatively unhazardous. Enzymatic digestion can be performed without a fume hood and with certainly less damage to microplastics. However, the efficacy of enzymes varies with the type of organic material present in the sample and is quite a time-consuming procedure. Every enzyme has its optimum pH and temperature that needs to be maintained and monitored continuously throughout the experiment (Miller et al. 2017; Stock et al. 2019). Enzymatic digestion can be carried out using cellulase, lipase, chitinase, protease, and proteinase-K (Prata et al. 2019; Stock et al. 2019). Proteinase K can digest >97% of the biological matter from the marine water samples without affecting microplastic particles (Cole et al. 2014). A limitation of enzymatic digestion is its high cost, especially when the sample size is very large. Enzyme protocols may include additional steps, such as first, the treatment of sediments with an industrial enzyme blend (2.5%) at 45°C for 60 min, then digestion with H_2O_2 (30%) to remove proteins and breakdown products from enzymatic treatment (Crichton et al. 2017; Prata et al. 2019).

2.4 IDENTIFICATION

The previous chapter discussed in depth that microplastics originate from variable sources, are composed of different polymers with a variety of shapes, sizes, and colors. Therefore, analytical identification is a critical part of microplastic research to ascertain the quantity and types of microplastics. Most commonly, visual inspection is done for the identification and quantification of microplastics, often combined with additional analytical examination for chemical characterization.

TABLE 2.3

Digestion methods for the removal of organic matter to improve the identification of microplastics, their efficiency, and effects on synthetic polymers.

Digestion	Treatment	Recovery rate	Polymer degradation	Organic matter degradation	Reference
Acid	HNO_3 (35%), 60°C 1 h	n.a.	Fusion of PET and HDPE; destruction of PA	100%	Catarino et al. (2017)
	HNO_3 (65%), RT overnight, 60°C 2 h, dilution 80°C distilled water	n.a.	PA degradation; yellowing	n.a.	Dehaut et al. (2016)
	HNO_3 (65%) and $HClO_4$ (65%) 4:1 overnight, boiled 10 min, dilution 80°C distilled water	n.a.	PA degradation; yellowing	n.a.	Dehaut et al. (2016)
	HNO_3 (5–69%), RT 96 h	<95%	Melted LDPE and PP; color change in PP, PVC, PET; decrease Raman peaks	n.a.	Karami et al. (2017)
	HNO_3 (55%) RT 1 month	n.a.	Whitening of PVC; degradation of PA	n.a.	Naidoo et al. (2017)
	HCl (5–37%), 25–60°C 96 h	n.a.	Changes in PET and PVC	>95%	Karami et al. (2017)
Alkali	NaOH, 60°C 1 h	94%	No	100%	Catarino et al. (2017)
	NaOH (10 M), 60°C 24 h	n.a.	CA degradation	n.a.	Dehaut et al. (2016)
	$K_2S_2O_8$ (0.27 M) and NaOH (0.24 M), 65°C 24 h	n.a.	CA degradation; unpredictable weight increase	n.a.	Dehaut et al. (2016)
	KOH (10%), RT 3 weeks	n.a.	No	n.a.	Dehaut et al. (2016)
	KOH (10%), 60°C 24 h	n.a.	CA degradation	n.a.	Dehaut et al. (2016)
	KOH (10%), 50°C 96 h	n.a.	Loss of PET and PVC	n.a.	Karami et al. (2017)
	KOH (10%), 40°C 96 h	n.a.	Loss of PET; yellowing of PA	n.a.	Karami et al. (2017)
	KOH (1 M), RT 2 days	n.a.	Degradation of LDPE, CA, CRADONYL, and PA	Most, except otoliths, squid beaks, paraffin, palm fat	Kühn et al. (2017)
	NaOH (1 mol L^{-1}), 17.5 ml of 65% HNO_3, and 2.5 ml ultrapure water and drying	95%	Degradation of PA, PET, EPS, LDPE, PVC; color change in PVC and PET	n.a.	Roch and Brinker (2017)

Oxidative	H_2O_2 (30%), 60°C for 1 h, 100°C for 7 h	n.a.	n.a.		Erni-Cassola et al. (2017)
	H_2O_2 (35%), RT, 40°C 96 h	n.a.	Decrease in Raman peaks of PVC and PA		Karami et al. (2017)
	H_2O_2 (35%), RT, 50°C 96 h	n.a.	Degradation of PA; color change of PET; foam and oxidization		Karami et al. (2017)
	H_2O_2 (6%) 70°C for 24 h	78% (PE)	n.a.		Sujathan et al. (2017)
	H_2O_2 (30%), 60°C until evaporation	n.a.	n.a.		Ziajahromi et al. (2017)
	NaClO (25–60°C)	n.a.	n.a.	<95%	Karami et al. (2017)
Enzymatic	Corolase 7086, 60°C 1 h	93%	No	n.a.	Catarino et al. (2017)
	Tripsin, 38–42°C 30 min	n.a.	No	88%	Courtene-Jones et al. (2017)
	Collagenase, 38–42°C 30 min	n.a.	No	76%	Courtene-Jones et al. (2017)
	Papain, 38–42°C 30 min	n.a.	No	72%	Courtene-Jones et al. (2017)
	Pepsin (0.5%) and HCL (0.063 M), 35°C 2 h, proteinase K (500 g ml⁻¹) and $CaCl_2$ 50°C 2 h, shaken 20 min, incubated 60°C 2 h, 30 ml H_2O_2 (30%) overnight	n.a.	No	Incomplete	Dehaut et al. (2016)
	15 ml Tris-HCl 60°C 60 min, proteinase K 60°C 2 h	97%	Calcium layer	n.a.	Karlsson et al. (2017)
	Proteinase-K (500 µg ml⁻¹, 50°C, 2 h. Followed by 5 M sodium perchlorate ($NaClO_4$) (RT, >20 minutes)	n.a.	No visible impact	>97%	Cole et al. (2014)

Source: Prata et al. (2019).

Note: CA—cellulose acetate; EPS—expanded polystyrene; HDPE—high-density polyethylene; LDPE—low-density polyethylene; n.a.—not available; Polymers: PA—polyamide (nylon); PE—polyethylene; PET—polyethylene terephthalate; PP—polypropylene; PS—polystyrene; PVC—polyvinyl chloride; RT—room temperature.

2.4.1 Visual Detection

Visual inspection is always the first step for screening of microplastics in any environmental samples. The presence of the microplastics can be detected using morphological characteristics and physical characteristics such as density. Further, visual identification is inexpensive and more practical to reduce the number of microplastics for subsequent chemical characterization. It is recommended to follow certain selection criteria to avoid discrepancy in the identification of microplastics (Hidalgo-Ruz et al. 2012; Wang and Wang 2018). Larger microplastics (>1 mm) can be visually recognized (i) if no cellular or organic structures are present, (ii) thickness of the fibers should be equal throughout the length, (iii) color must be homogeneous and clear; if it is transparent or white, they must be re-examined under high magnification and/or a fluorescence microscope. The weathering of microplastic particles alters morphology, thus visual identification becomes more difficult. The plastics fragments may be confused with biological or inorganic matter leading to overestimation. Subsequently, chemical analysis of the sorted microplastics often confirms them to be non-plastic (Lv et al. 2021). Use of staining dyes to aid visual identification is a low-cost approach. Nile Red appears to be a promising stain for microplastic fragments, having a short incubation time (10–30 min), good recovery rates (96.6%), and further vibrational spectroscopy can be performed with or without short bleaching (Prata et al. 2019; Lv et al. 2021). In addition, semi-automated techniques (ZooScan, flow cytometry, cell sorters, coulter counters) can be used to decrease the time while analyzing a large number of samples (Lusher et al. 2017). However, these equipments are costly and sophisticated and requires technical expertise. Scanning electron microscopes (SEMs) produce high-resolution images by shooting a high-intensity electron beam at the sample surface and scanning it in a raster scan pattern. SEMs have been used in a lot of studies, since in high-resolution images it is comparatively easier to differentiate microplastics from other organic or inorganic impurities due to different surface morphology. Transmission electron microscopy (TEM) can also be used for the identification of microplastic fragments (Woo et al. 2021). Further, combining an EDS (energy dispersive spectroscopy) detector to SEM provides additional information about the elemental composition of microplastics containing inorganic additives. For instance, additives in microplastics, such as Al, Ca, Mg, Na, and Si, or antioxidants, can be detected as markers and used for identification using EDS (Watteau et al. 2018). Elemental analysis allows differentiation of natural materials from plastics fragments, which thereby narrows the amount of particles required for subsequent spectroscopic analysis. Visual identification is a subjective approach that produces variable results with different observers. For example, multiple observers had a detections range of 60–100% while identifying microplastics directly in beach sediments; variation could be due to individual experience, fatigue, etc. (Lavers et al. 2016).

2.4.2 Spectroscopic Detection

Fourier transform infrared (FTIR) spectroscopy can produce a distinct infrared spectrum for the chemical bond. Since different materials have distinct bond compositions, any unknown material can be identified by comparing its spectrum to the spectra of known materials. FTIR is one of the most widely used analytical methods for the chemical characterization of the microplastics recovered from water, sediment, and biota. FTIR provides consistent and accurate results rapidly, and is not affected by the fluorescence interference. In addition to the identification of polymer types, FTIR also gives further information about physiochemical weathering of microplastics by assessing oxidation intensity (Wang and Wang 2018). FTIR spectrum can give information about the oxidation state of microplastic particles using composition of oxygen bonds such as carbonyl groups (Zhou et al. 2020). ATR (Attenuated total reflection) FTIR can identify microplastics of irregular shape larger than 500 μm. Focal plane array (FPA) detector-based micro-FTIR facilitates concurrent generation and recording of large numbers of chemical spectra in a single measurement of even an entire filter (Zarfl 2019). This particularly is very useful in reducing imaging time from many days to less than 9 h. FPA-based micro-FTIR can identify microplastics of much smaller sizes (Lv et al. 2021). FTIR microscopes can provide spatial resolutions to 5 μm, minimum required sample thickness of approximately 150 nm analyzed (Song et al. 2015; Wang and Wang 2018; Nguyen et al. 2019). FTIR is more suitable to determine polymeric composition of microplastic particles larger than 20 μm; plastics smaller than the aperture size can be missed (Song et al. 2015; Wang and Wang 2018; Nguyen et al. 2019). The only limitation is the huge cost of the instrument and a highly skilled technician is required for operating the instrument.

Raman spectroscopy is another frequently used and extremely accurate approach for polymer characterization of microplastics from various environmental samples. Raman spectroscopy excites target samples using a laser, and similarly to IR spectroscopy, captures the spectrum of the induced vibration from the sample. Combining Raman spectroscopy with a microscope can facilitate identification of microplastic particles of variable sizes. Raman microscopy using a laser beam can focus on a smaller area than FT-IR, thus allowing analysis of the microplastic fractions less than 20 μm in size. Raman spectroscopy of microplastics is usually performed to obtain specific information on high molecular weight polymers (Gong and Xie 2020). This technique offers an advantage to localize polymer particles in biological tissues up to subcellular levels. However, Raman microscopy is not suitable for microplastic particles that are dyed, especially dark colors, since signals from dye interfere with the signal from polymers, and dark colors can cause fluorescence or burning (Miller et al. 2017). Further, presence of pigments or photosensitive materials in the samples also interfere with the identification of the polymers (Gong and Xie 2020). For both FTIR and Raman, there is minimal requirement of sample preparation, besides cleanup for FTIR. However, the former is destructive, while the latter is non-destructive, since the sample does not need to be manipulated or flattened.

2.4.3 Thermal Detection

Pyrolysis gas chromatography/mass spectrometry (Pyr-GC-MS) is also successfully employed for microplastic analysis. Pyr-GC-MS is a destructive technique that allows direct analysis of the unknown samples with minimal pretreatment. The method identifies polymers by analyzing their thermal degradation products. Briefly, microplastics are pyrolyzed (thermally decomposed) under inert conditions and the generated gas is cryo-trapped, subsequently separated on a chromatographic column. It generates Pyrolysis GC chromatograms of the samples that can be compared with the reference chromatograms of the known pure polymers. The advantage of Pyr-GC-MS is that it provides detailed information about the chemical composition of the polymer and contained organic additives (Wang and Wang 2018). However, it cannot differentiate size and shape. Decomposition products of different polymers could be similar, which may result in misinterpretation. Further, only a small amount of sample can be analyzed at a time; a large number of samples may require a lot of time for analysis (Gong and Xie 2020). Thermal analysis is a potential substitute to spectroscopy for the chemical characterization of specific polymer types. The only limitation is that further analysis of samples is not possible because it is destroyed during analysis.

2.4.4   EMERGING IDENTIFICATION METHODS

The elemental composition of polymers can be detected with an X-ray fluorescence (XRF) spectrometer using diffraction and reflection of radiation, further potentially identifying some additives or adsorbed metals (Prata et al. 2019). Bulk microplastics fraction can be separated from soil/sediment samples through pressurized fluid extraction with solvents such as hexane, methanol, and dichloromethane under optimum temperature and pressure conditions (Prata et al. 2019). AFM coupled with IR or Raman spectroscopy could be a potential tool to study microplastics and further nanoplastics. AFM probes may work in both contact and non-contact modes. It can generate nanoscale resolution images. Furthermore, combining with IR or Raman spectroscopy can give additional information about the chemical structure of the study material (Woo et al. 2021). Nano-thermal analysis (nano-TA) is another promising technique that produces AFM images using nano-TA probes. It is a local thermal analysis technique combined with high spatial resolution. Through this, an understanding of the thermal behavior of materials can be obtained with a spatial resolution of less than 100 nm. In nano-TA analysis, the probe moves to a fixed point on the sample surface when a point of interest is selected. The probe is very sensitive to the stiffness (hardness) of the microplastic (Woo et al. 2021). Thermal gravimetric analysis-Fourier transform infrared spectroscopy-gas chromatography-mass spectrometry (TGA-FTIR-GC-MS) allows microplastics monitoring in real time and facilitates both qualitative and quantitative analyses, in contrast to the current TGA-FTIR that identifies in real time but provides only qualitative data. By adding a mass spectrometry step to TGA-FTIR-GC-MS it allows precise evaluation of the thermal decomposition products; thus different polymer types such as PE, PP, PS, PVC can be identified accurately (Liu et al. 2021). See Table 2.4 for common methods of analyzing microplastics.

2.5   ASSURING QUALITY IN MICROPLASTIC MONITORING

When conducting microplastics monitoring programs, it is critical to implement strict quality assurance and quality control methods throughout the complete process in order to improve data quality. Major barriers in detecting microplastics retrieved

TABLE 2.4

Common methods of analyzing microplastics from the environment.

Analysis	Definition	Advantage	Disadvantage	References
Microscopy	Analyze substances using a microscope. Optical microscopes to electron microscopes such as SEM and TEM are widely used.	Able to directly observe the surface Simple sample preparation	Takes the longest time to determine Depending on the researcher, the result judgment may vary, resulting in low accuracy	Watteau et al. (2018); Morgado et al. (2021)
Spectroscopy	Analyze the surface of a material using optical technology	Multiple samples can be analyzed at once, and the method preserves the sample Existing data library can be used for analysis	Difficult to discriminate large amounts of samples Takes a long time to identify	Kniggendorf et al. (2019); Zhou et al. (2020)
Thermal analysis	Analyze using heat based on the unique physical and chemical properties of each substance	A large number of samples can be identified at once Existing data library can be used for analysis	Samples are not preserved Quantitative analysis of each sample is not possible	Zainuddin and Syuhada (2020); Liu et al. (2021)
Emerging Technology	Analyze using a new technology that is attracting attention instead of the previously described analysis methods	Development in the direction of complementing the shortcomings of existing technologies	Less data accumulated than existing technologies	Erni-Cassola et al. (2017); Garaba and Dierssen (2018)

Source: Woo et al. (2021).

from the marine environment are primarily contamination and under-/over-estimation. There are some recommended precautions to follow to avoid sample contamination during analysis of microplastics and to ensure processing quality and efficacy.

1. Avoid external plastics as much as you can. Plastic equipment and plasticware should be avoided or replaced with respective objects made from alternative materials such as glass or aluminum.
2. Clothes worn while sample collecting and processing in the laboratory should not be of synthetic fibers; it is better to wear clothes made of cotton.
3. To minimize cross-contamination of the environmental sample with possible plastic particles, all chemicals and solutions used for sample processing should be pre-filtered prior to use and all open beakers and jars should be covered.

4. Airborne contamination can be checked installing dust boxes and air filter units in the lab. If the sample is too small, additional technical measures can be adopted such as filtering under laminar flow cabinets.
5. Samples should be processed in replicates for higher reliability. During the sampling and processing of the sample, a clean membrane can be included as an environmental control, and must be analyzed with the same detection process as the sample.
6. Including procedural blanks/controls is highly recommended while running samples for chemical characterization using FTIR, GC-MS, etc. to observe contamination during processing of the samples. Additionally, spiked blanks can be included to check the recovery rate.
7. Before doing the analysis of the environmental samples, protocols must be standardized for microplastic analysis.
8. Prior to using chemical digestants for sample digestion, it is critical to conduct a comprehensive test to measure the potential impacts of the applied chemical digestants on plastics.

2.6 SUMMARY

For better understanding and limiting the effects of microplastics, monitoring and evaluating their occurrence and distribution across various environments including marine is required. Standardized processes and approaches are essential to establish uniformity in the data of different samples and locations worldwide. Sampling is the first and foremost step to study microplastics distribution and occurrence. Sampling should be done with utmost care to ensure precise estimation and identification, and eliminate external contamination. Sample size is important to ensure accurate monitoring. For suspended microplastic particles in marine surface water, bongo/zooplankton nets, plankton samplers, neuston nets, and manta trawls are commonly used while for the shallow bottom water and sediments, bottom trawls can be used. Selection of the mesh size may depend on the target particle size. Marine organisms were found to ingest microplastics; therefore it is important to monitor microplastic uptake in the marine organism and assess their impact. Nektonic and planktonic animals and invertebrates can be collected using manta and bongo nets.

Preliminary screening of the samples can be done visually on the site itself, and the screened microplastics can be brought to a laboratory for further examination. The collected microplastics must be separated from the samples in order to get clean particles for subsequent quantification and identification. Filtration or sieving, density separation, and digestive processes are the usual separation procedures. Size-based sorting can be done using sieves or filters of different mess or pore size respectively. Density-based separation microplastics can be separated by flotation with saturated salt solutions of considerably high density followed by filtration of the supernatant. In an electrostatic separator, the conductive sediments and other particulate matter can be separated from non-conductive microplastics based on the difference in conductive properties.

The microplastics ingested by the marine biota can be isolated and identified after dissection, homogenization, depuration, and digestion of the organism tissues with chemicals or enzymes. Microplastics are commonly investigated in the stomach or the gastrointestinal tract of the bigger organisms while smaller organisms can be investigated entirely. Acid or alkali digestion uses acids or alkali respectively in different concentrations and at different temperatures for variable duration to remove the organic matter from the samples. Increasing the molarity and the temperature could enhance digestion. However, it could also degrade polymers. Oxidizing agents can also be used for removing organic matter. Enzymatic digestion is considered as an alternative to chemical digestion and comparatively unhazardous. Enzymatic digestion can be carried out using cellulase, lipase, chitinase, protease, proteinase-K without affecting microplastic particles; however, it could be costly for a large number samples.

Analytical identification is a critical part of microplastic research to ascertain quantity and types of microplastics. Most commonly, visual inspection is done for the identification and quantification of microplastics. Samples that are difficult to study through the naked eye can be observed under microscopes, further using scanning electron microscopes (SEM) and transmission electron microscopy (TEM). Additional analytical examination can be done for chemical characterization, using SEM-EDS (energy dispersive spectroscopy), Fourier transform infrared (FTIR) spectroscopy, Raman spectroscopy, Pyrolysis gas chromatography/mass spectrometry (Pyr-GC-MS), AFM coupled with IR or Raman spectroscopy, nano-thermal analysis (nano-TA), etc.

When conducting microplastics monitoring programs, it is critical to implement strict quality assurance and quality control methods throughout the complete process in order to improve data quality. Major barriers in detecting microplastics retrieved from the marine environment are primarily contamination and under-/over-estimation. External contamination must be avoided during sampling and processing. Procedural blanks/controls are highly recommended.

REFERENCES

Catarino, A. I., Thompson, R., Sanderson, W. and Henry, T. B. 2017. Development and optimization of a standard method for extraction of microplastics in mussels by enzyme digestion of soft tissues. *Environmental Toxicology and Chemistry*, 36(4):947–951.

Cole, M., Webb, H., Lindeque, P. K., Fileman, E. S., Halsband, C. and Galloway, T. S. 2014. Isolation of microplastics in biota-rich seawater samples and marine organisms. *Scientific Reports*, 4:1–8.

Courtene-Jones, W., Quinn, B., Murphy, F., Gary, S. F. and Narayanaswamy, B. E. 2017. Optimisation of enzymatic digestion and validation of specimen preservation methods for the analysis of ingested microplastics. *Analytical Methods*, 9(9):1437–1445.

Crichton, E. M., Noël, M., Gies, E. A. and Ross, P. S. 2017. A novel, density-independent and FTIR-compatible approach for the rapid extraction of microplastics from aquatic sediments. *Analytical Methods*, 9(9):1419–1428.

Cutroneo, L., Reboa, A., Geneselli, I. and Capello, M. 2021. Considerations on salts used for density separation in the extraction of microplastics from sediments. *Marine Pollution Bulletin*, 166:112216.

Dehaut, A., Cassone, A. L., Frère, L., Hermabessiere, L., Himber, C., Rinnert, E., Rivière, G., Lambert, C., Soudant, P., Huvet, A., Duflos, G. and Paul-Pont, I. 2016. Microplastics in seafood: Benchmark protocol for their extraction and characterization. *Environmental Pollution*, 215:223–233.

Elizalde-Velázquez, G. A. and Gómez-Oliván, L. M. 2021. Microplastics in aquatic environments: A review on occurrence, distribution, toxic effects, and implications for human health. *Science of The Total Environment*, 780:146551.

Elsayed, A. A., Erfan, M., Sabry, Y. M., Dris, R., Gaspéri, J., Barbier, J. S., Marty, F., Bouanis, F., Luo, S., Nguyen, B. T. T., Liu, A. Q., Tassin, B. and Bourouina, T. 2021. A microfluidic chip enables fast analysis of water microplastics by optical spectroscopy. *Scientific Reports*, 11(1):1–11.

Erni-Cassola, G., Gibson, M. I., Thompson, R. C. and Christie-Oleza, J. A. 2017. Lost, but found with Nile red: A novel method for detecting and quantifying small microplastics (1 mm to 20 μm) in environmental samples. *Environmental Science and Technology*, 51(23):13641–13648.

Felsing, S., Kochleus, C., Buchinger, S., Brennholt, N., Stock, F. and Reifferscheid, G. 2018. A new approach in separating microplastics from environmental samples based on their electrostatic behavior. *Environmental Pollution*, 234:20–28.

Garaba, S. P. and Dierssen, H. M. 2018. An airborne remote sensing case study of synthetic hydrocarbon detection using short wave infrared absorption features identified from marine-harvested macro- and microplastics. *Remote Sensing of Environment*, 205:224–235.

Gong, J. and Xie, P. 2020. Research progress in sources, analytical methods, eco-environmental effects, and control measures of microplastics. *Chemosphere*, 254: 126790.

Herrera, A., Garrido-Amador, P., Martínez, I., Samper, M. D., López-Martínez, J., Gómez, M. and Packard, T. T. 2018. Novel methodology to isolate microplastics from vegetal-rich samples. *Marine Pollution Bulletin*, 129(1):61–69.

Hidalgo-Ruz, V., Gutow, L., Thompson, R. C. and Thiel, M. 2012. Microplastics in the marine environment: A review of the methods used for identification and quantification. *Environmental Science and Technology*, 46(6):3060–3075.

Imhof, H. K., Schmid, J., Niessner, R., Ivleva, N. P. and Laforsch, C. 2012. A novel, highly efficient method for the separation and quantification of plastic particles in sediments of aquatic environments. *Limnology and Oceanography: Methods*, 10(7):524–537.

Karami, A., Golieskardi, A., Choo, C. K., Romano, N., Ho, Y. bin and Salamatinia, B. 2017. A high-performance protocol for extraction of microplastics in fish. *Science of the Total Environment*, 578:485–494.

Karlsson, T. M., Vethaak, A. D., Almroth, B. C., Ariese, F., van Velzen, M., Hassellöv, M. and Leslie, H. A. 2017. Screening for microplastics in sediment, water, marine invertebrates and fish: Method development and microplastic accumulation. *Marine Pollution Bulletin*, 122(1–2):403–408.

Kniggendorf, A. K., Wetzel, C. and Roth, B. 2019. Microplastics detection in streaming tap water with Raman spectroscopy. *Sensors*, 19(8):1839.

Kühn, S., van Werven, B., van Oyen, A., Meijboom, A., Bravo Rebolledo, E. L. and van Franeker, J. A. 2017. The use of potassium hydroxide (KOH) solution as a suitable approach to isolate plastics ingested by marine organisms. *Marine Pollution Bulletin*, 115(1–2):86–90.

Lavers, J. L., Oppel, S. and Bond, A. L. 2016. Factors influencing the detection of beach plastic debris. *Marine Environmental Research*, 119:245–251.

Liu, M., Lu, S., Chen, Y., Cao, C., Bigalke, M. and He, D. 2020. Analytical methods for microplastics in environments: Current advances and challenges. In: *Handbook of Environmental Chemistry*. New York: Springer Science and Business Media Deutschland GmbH, 3–24.

Liu, Y., Li, R., Yu, J., Ni, F., Sheng, Y., Scircle, A., Cizdziel, J. V. and Zhou, Y. 2021. Separation and identification of microplastics in marine organisms by TGA-FTIR-GC/MS: A case study of mussels from coastal China. *Environmental Pollution*, 272.

Löder, M. G. J. and Gerdts, G. 2015. Methodology used for the detection and identification of microplastics—A critical appraisal. *Marine Anthropogenic Litter*, 201–227.

Lusher, A. L., Welden, N. A., Sobral, P. and Cole, M. 2017. Sampling, isolating and identifying microplastics ingested by fish and invertebrates. *Analytical Methods*, 9(9):1346–1360.

Lv, L., Yan, X., Feng, L., Jiang, S., Lu, Z., Xie, H., Sun, S., Chen, J. and Li, C. 2021. Challenge for the detection of microplastics in the environment. *Water Environment Research*, 93(1):5–15.

Mai, L., Bao, L. J., Shi, L., Wong, C. S. and Zeng, E. Y. 2018. A review of methods for measuring microplastics in aquatic environments. *Environmental Science and Pollution Research*, 25(12):11319–11332.

Malankowska, M., Echaide-Gorriz, C. and Coronas, J. 2021. Microplastics in marine environment: A review on sources, classification, and potential remediation by membrane technology. *Environmental Science: Water Research and Technology*, 7(2):243–258.

Miller, M. E., Kroon, F. J. and Motti, C. A. 2017. Recovering microplastics from marine samples: A review of current practices. *Marine Pollution Bulletin*, 123(1–2), 6–18.

Morgado, V., Palma, C. and Bettencourt da Silva, R. J. N. 2021. Microplastics identification by infrared spectroscopy—Evaluation of identification criteria and uncertainty by the Bootstrap method. *Talanta*, 224.

Naidoo, T., Goordiyal, K. and Glassom, D. 2017. Are nitric acid (HNO_3) digestions efficient in isolating microplastics from juvenile fish? *Water, Air, and Soil Pollution*, 228(12).

Nguyen, B., Claveau-Mallet, D., Hernandez, L. M., Xu, E. G., Farner, J. M. and Tufenkji, N. 2019. Separation and analysis of microplastics and nanoplastics in complex environmental samples. *Accounts of Chemical Research*, 52(4):858–866.

Nuelle, M. T., Dekiff, J. H., Remy, D. and Fries, E. 2014. A new analytical approach for monitoring microplastics in marine sediments. *Environmental Pollution*, 184:161–169.

Pagter, E., Frias, J. and Nash, R. 2018. Microplastics in Galway Bay: A comparison of sampling and separation methods. *Marine Pollution Bulletin*, 135:932–940.

Phuong, N. N., Fauvelle, V., Grenz, C., Ourgaud, M., Schmidt, N., Strady, E. and Sempéré, R. 2021. Highlights from a review of microplastics in marine sediments. *Science of The Total Environment*, 777:146225.

Prata, J. C., da Costa, J. P., Duarte, A. C. and Rocha-Santos, T. 2019. Methods for sampling and detection of microplastics in water and sediment: A critical review. *TrAC—Trends in Analytical Chemistry*, 110:150–159.

Roch, S. and Brinker, A. 2017. Rapid and efficient method for the detection of microplastic in the gastrointestinal tract of fishes. *Environmental Science and Technology*, 51(8):4522–4530.

Sgier, L., Freimann, R., Zupanic, A. and Kroll, A. 2016. Flow cytometry combined with viSNE for the analysis of microbial biofilms and detection of microplastics. *Nature Communications*, 7(1):1–10.

Shimizu, K., Sokolov, S. v., Kätelhön, E., Holter, J., Young, N. P. and Compton, R. G. 2017. In situ detection of microplastics: Single microparticle-electrode impacts. *Electroanalysis*, 29(10):2200–2207.

Song, Y. K., Hong, S. H., Jang, M., Han, G. M., Rani, M., Lee, J. and Shim, W. J. 2015. A comparison of microscopic and spectroscopic identification methods for analysis of microplastics in environmental samples. *Marine Pollution Bulletin*, 93(1–2):202–209.

Stock, F., Kochleus, C., Bänsch-Baltruschat, B., Brennholt, N. and Reifferscheid, G. 2019. Sampling techniques and preparation methods for microplastic analyses in the aquatic environment—A review. *TrAC Trends in Analytical Chemistry*, 113:84–92.

Sujathan, S., Kniggendorf, A. K., Kumar, A., Roth, B., Rosenwinkel, K. H. and Nogueira, R. 2017. Heat and bleach: A cost-efficient method for extracting microplastics from return activated sludge. *Archives of Environmental Contamination and Toxicology*, 73(4):641–648.

van Cauwenberghe, L., Vanreusel, A., Mees, J. and Janssen, C. R. 2013. Microplastic pollution in deep-sea sediments. *Environmental Pollution*, 182:495–499.

Wang, W. and Wang, J. 2018. Investigation of microplastics in aquatic environments: An overview of the methods used, from field sampling to laboratory analysis. *TrAC Trends in Analytical Chemistry*, 108:195–202.

Watteau, F., Dignac, M. F., Bouchard, A., Revallier, A. and Houot, S. 2018. Microplastic detection in soil amended with municipal solid waste composts as revealed by transmission electronic microscopy and pyrolysis/GC/MS. *Frontiers in Sustainable Food Systems*, 2.

Woo, H., Seo, K., Choi, Y., Kim, J., Tanaka, M., Lee, K. H. and Choi, J. 2021. Methods of analyzing microsized plastics in the environment. *Applied Sciences*, 211(22):10640.

Zainuddin, Z. and Syuhada. 2020. Study of analysis method on microplastic identification in bottled drinking water. *Macromolecular Symposia*, 391(1).

Zarfl, C. 2019. Promising techniques and open challenges for microplastic identification and quantification in environmental matrices. *Analytical and Bioanalytical Chemistry*, 411(17):3743–3756.

Zhou, L., Wang, T., Qu, G., Jia, H. and Zhu, L. 2020. Probing the aging processes and mechanisms of microplastic under simulated multiple actions generated by discharge plasma. *Journal of Hazardous Materials*, 398.

Ziajahromi, S., Neale, P. A., Rintoul, L. and Leusch, F. D. L. 2017. Wastewater treatment plants as a pathway for microplastics: Development of a new approach to sample wastewater-based microplastics. *Water Research*, 112:93–99.

3 Ecological Risks and Environmental Fate of Microplastic Pollution

3.1 INTRODUCTION

Microplastic pollution in recent years has spurred a wide range of global concerns due to its ubiquity, serious implications for ecosystems, and potential hazards for human health. Presence of a pollutant, or of a foreign element, in an environment is never a neutral one; it is, along with the physical environment it becomes a part of, in a relationship with the biotic component of the environment. As every element in an environment has its own fate associated with its entire life cycle so do microplastics. Plastics fragmented into microplastics strengthen their "bond" with living organisms and become part of the food chain and food web.

When, through a food chain, the microplastics enter into organisms' bodies, they would affect the body systems in different ways, including imparting their toxic effects. Since the marine environment has become a kind of "natural pool" of the microplastic pollutants, their large-scale impact on marine ecosystems—physical aquatic environment as well as marine organisms—would naturally be of a higher degree and greater proportions.

3.2 PLASTICS USED IN THE MARINE ENVIRONMENT

Almost all plastic products are used in human habitats on the land of the Earth. The debris that accumulates in the oceans is all because of anthropogenic reasons and accounts for as much as 75% of land-based sources (Andrady 2011). Since there is interconnection between terrestrial and aquatic ecosystems, the wastes/pollutants generated in terrestrial ecosystems (human settlements) would link up with the aquatic ecosystems and vice versa. As the marine waters are at the lowest locations, and all streams, rivers, and rivulets originating in the mountains, hills, or uplands end their journey into oceans, the plastic wastes, once having reached waterways, ultimately dump into oceans where they keep on accumulating and where their subsequent fate is determined. Easily transported by means of winds and water currents (Tushari and Senevirathna 2020), the used-up/discarded plastic material gets easy accessibility to the ocean's environments. Microplastics' unique characteristics, like their high degree of non-biodegradability and long life, are favorable for them persisting in the marine environment for a very long time.

Plastic debris floating over the surface of the globe's oceans accounts for more than 5 trillion plastic pieces and 260,000 tons in weight (Eriksen et al. 2014) and

DOI: 10.1201/9781003312086-3

is distributed irrespective of the developing or developed regions (Tushari and Senevirathna 2020). Wastes generated from both primary and secondary plastic sources are accumulating in ocean waters. Tourism and recreational activities are one of the significant human activities responsible for plastic debris accumulation in the coastal and ocean ecosystems.

3.3 MICROPLASTICS IN THE MARINE ECOSYSTEM

The plastic materials that accumulate in the marine waters are of four categories based on their sizes: (i) megaplastics (>1 m), (ii) macroplastics (<1 m), (iii) mesoplastics (<2.5 cm), and (iv) microplastics (<5 mm). Among these plastics, the microplastics originate *in situ* as a result of the fragmentation of mega-, macro-, and mesoplastics or directly enter into oceans along with plastic waste comprising cosmetic and personal care materials. Fragmentation of larger plastic particles increasing the number of microplastic particles in the oceans may be due to physical, chemical, and biological factors (Wang et al. 2018) operating in the ocean ecosystem. Overall, microplastics' fate in the ocean environment as per the current understanding (Wayman and Niemann 2021) is shown in Figure 3.1.

While the microplastics are distributed in the water column in the marine and coastal systems according to their density, the synthetic pollutants released from the plastics interact with metals and organic pollutants (Guo and Wang 2019). Plastics debris comprises the plastic materials of varying density, a property that also determines the distribution behavior of plastic particles. For example, low-density particles

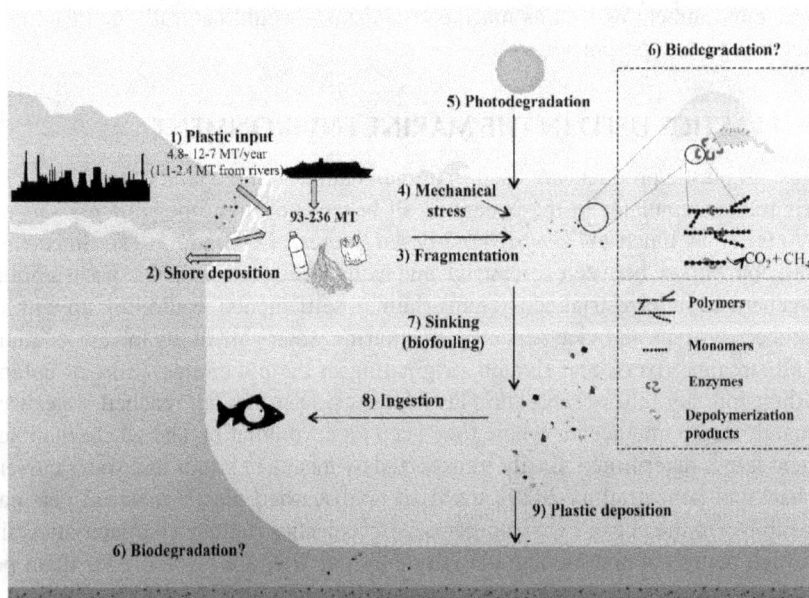

FIGURE 3.1 The fate of plastics in the ocean environment.

Source: Wayman and Niemann (2021).

like polyethylene (PE) and polypropylene (PP) float on the water surface, whereas high-density particles of polystyrene (PS), polyamide (PA), polyvinyl chloride (PVC), and polyethylene terephthalate (PET) get deposited towards the lower level of the water column (Guo and Wang 2019). In this way, microplastic pollutants tend to distribute in every sub-zone/layer—pelagic and benthic—of the coastal and marine ecosystems. Microplastics and nanoplastics are not there only in the physical environment of the marine ecosystem, but are also found in the bodies of the organisms.

3.4   ECOLOGICAL RISKS OF THE MICROPLASTICS TO MARINE LIFE

Marine environment represents the Earth's largest habitat for numerous species belonging to all the five living kingdoms. Ocean life is believed to be more primitive. Most of the life in the marine environment is fed by photosynthesis, but communities based exclusively on chemosynthesis around hydrothermal vents on the ocean floor also flourish (Singh 2019, 2020). Open oceans also represent the most stable ecosystems on the planet. Again, oceans play a key role in maintaining the thermodynamics of the biosphere and regulating the planet's climate patterns.

3.4.1   Physical Effects of Microplastics to Marine Organisms

Fragmentation of mega-, macro-, and mesoplastics into marine environments results in the widespread distribution of microplastics in the marine environment. Plastics being forcefully bombarded by water releases microplastics from larger plastic. The microplastics then spread horizontally across the ocean surface as well as vertically across the water column, depending on many factors relating to the ocean environment and the plastic fragmentation processes.

Microplastics as foreign particles or pollutants change the marine environment the same way as air pollutants change the atmosphere. Changes in the physical environment as a result of the presence of microplastics induce alterations in ecological processes, making them much more unfavorable for marine organisms. Changes in the ecological processes due to microplastics' physical effects on the marine ecosystem would largely depend on microplastic concentrations. The higher the microplastics concentrations the greater the magnitude of microplastic-induced changes.

The biology of aquatic environments corresponds broadly to variations in physical factors such as light, temperature, and water movements and to chemical factors such as salinity and oxygen (Molles 2005). Increasing microplastic concentrations would potentially alter the physical factors that, in turn, would phenomenally influence marine life.

About 80% of the light striking the ocean surface is absorbed in the first 10 m. Ultraviolet and infrared light is absorbed in the first few meters. The plastic debris and microplastics floating on the ocean surface are likely to reduce the light absorption capacity because of two reasons: (i) plastic particles directly obstruct the light striking the ocean surface or reflect back the light, and (ii) the various additives and other colors and chemicals would reduce water clarity/transparency, thus reducing light absorption. Availability of insufficient light will directly hit photosynthetic efficiency of the marine ecosystem, leading eventually to reduced primary and, consequently, secondary productivity and overall ecosystem functioning. Microplastics and larger plastic particles may also shield photosynthesizing phytoplankton, resulting in the curtailment of the photosynthesis.

Oxygen level in ocean waters is much less (about 9 ml per liter of water) than in the air (200 ml per liter of air). Oxygen concentration is maximum at the ocean surface and progressively decreases with depth, attaining a minimum level at around 1000 m depth. Plastic debris and microplastic particles further contribute to reducing oxygen exchange in marine ecosystems. Microorganisms proliferating in the "plastisphere" within the hydrosphere also contribute to reducing marine oxygen concentrations by consuming greater amounts of oxygen during their decomposition processes. Reduced levels of dissolved oxygen result in declining metabolic rates and depressed energetics of living organisms, thus putting marine life under a kind of environmental stress.

Marine environments constitute the largest carbon sink. Although CO_2 sequestration efficiency of the oceans depends on many atmospheric and aquatic factors, plastic debris and microplastic pollution may significantly reduce carbon absorption efficiency of the marine ecosystems, which would have serious implications on atmospheric carbon balance and exert serious impacts on marine as well as on terrestrial life.

Salinity (i.e., amount of salt dissolved in water) of the oceans varies with latitude and among the seas that fringe the oceans. In the open ocean the salinity varies from about 34 to 36.5 g per kg of water (Molles 2005). Although microplastic pollution is unlikely to significantly affect ocean salinity, in some pockets the presence of extremely high microplastic concentrations may impact the salinity. The salinity, on the other hand, is a critical factor determining chemical degradation of the plastics. Thus, marine and coastal environments comprising about 0.5 to 35 g of salt per kg of seawater are highly susceptible to microplastic formation. The overall impact of microplastics on marine life is through their interaction with other operating factors rather than their influence on salinity.

Changes in ocean temperatures due to microplastic pollution do not seem to be an issue. However, it is very likely that the microplastic particles, along with floating mega-, macro-, and meso-particles of the plastics, would obstruct loss of heat absorbed as light and the heat generated within the ecosystem through metabolic processes to the atmosphere, thereby contributing to an increase in internal temperature (see Figure 3.2).

3.4.2 Toxicity of Ingested Microplastics

The ubiquitous nature of microplastics makes it almost impossible for human beings to save themselves from their potential effects on health. Microplastics frequently enter the human body through inhalation and ingestion of water and foods. Upon entering the human body in higher concentrations and under some specific conditions, the microplastics can induce physical and chemical toxicity (SAPEA 2019). Risks of microplastic-induced toxicity are greater to the people and communities living in the proximity of plastic-producing factories or to those dependent on or fond of seafood.

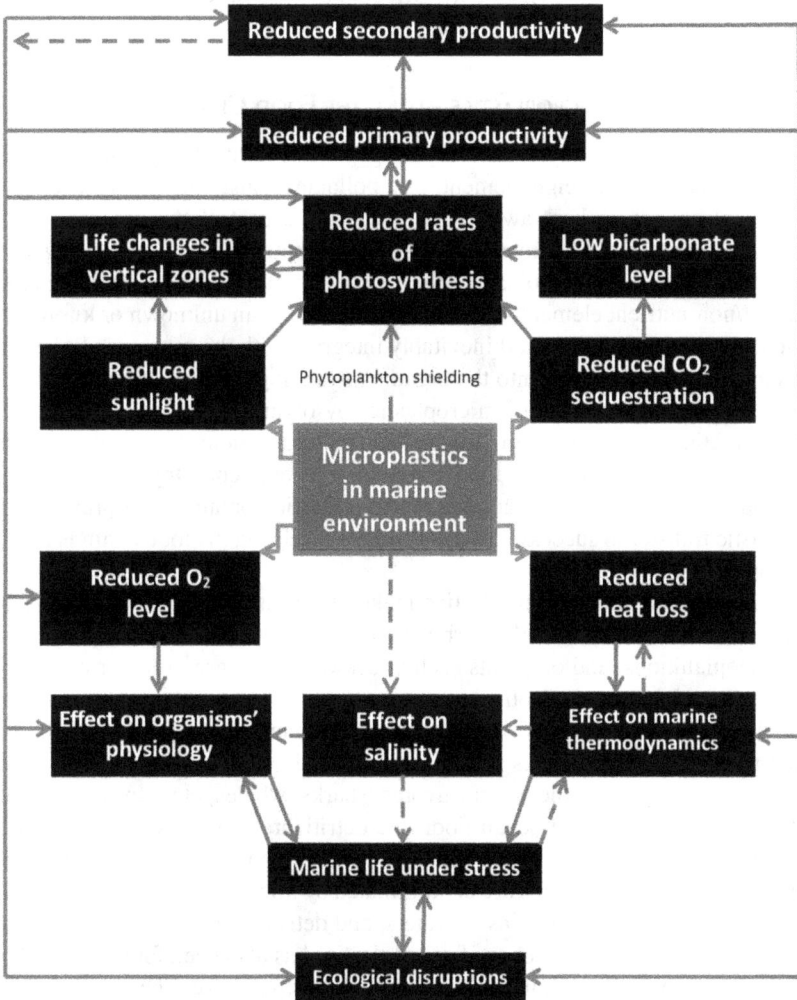

FIGURE 3.2 Ecological risks associated with microplastic pollution in the marine environment (solid arrows depict direct impact, broken ones show weak or indirect impact).

Major toxicity effects attributable to the ingested microplastics are owing to: (i) the leakage in the respiratory and digestive tracts of chemical additives and plastic quality-enhancing chemicals (e.g., plasticizers, light and heat stabilizers, antioxidants, flame retardants, etc.) added during the plastic production processes; and (ii) microplastic-associated contaminants, such as persistent organic pollutants (POPs), heavy metals, etc. Thousands of chemicals, including toxic ones (e.g., endocrine-disrupting chemicals), spill out their toxic effects on the human body resulting in, for instance, obesity, diabetes, heart attack, infertility, sex malformation, cancer, behavioral changes, neurological disorders, etc. (Barrett 2019; Kannan and

Vimalkumar 2021). Such serious toxic effects of microplastic ingestion have been discussed in detail in Chapter 4.

3.4.3 Transfer of Microplastics Along the Food Chain

The first law of ecology says that everything is related with everything else. Thus, any contaminant, or a foreign element, or a pollutant registering its presence in the environment cannot be kept away from becoming a part of the processes linked with living organisms. Organisms subsist on the nutrient sources existing in the environment. If the sources of nutrients become contaminated by an undesirable/ unwanted/non-nutrient element, chemical compound, or an unknown or known toxicant in the environment, it would inevitably integrate with the processes linked with nourishment and make entry into the organisms ranging from autotrophs to top carnivores. Accidental ingestion of microplastics by organisms is common. However, exposure of the organisms to microplastics becomes a natural phenomenon when the microplastics enter into the food chain of an ecosystem. Organisms' exposure to microplastics through the ingestion of microplastic-contaminated prey and thus microplastic transfer to successive higher trophic levels via the food chain is referred to as trophic transfer.

In marine ecosystems, microplastics make entry into the first order consumers (the second trophic level, i.e., the herbivores), when they feed on phytoplankton (in case of zooplanktons) and on plants such as seaweeds (in case of larger herbivores). When the zooplanktons and other herbivores in the marine ecosystems are eaten by the first order carnivores or second order consumers (the organisms at the third trophic level), the microplastics pass on to their bodies. In this way, the microplastics make their way up to the top carnivores (sharks, whales, etc.). In the organisms inhabiting sediments on the ocean floor (the detritivores), the microplastics will be transferred through detritus food chains in the same manner. Thus, all the trophic levels in the marine ecosystems are contaminated by microplastics through operating food chains—grazing food chains in waters, and detritus food chains in the organic matter in sediments. The presence of microplastics has also been found in the digestive tracts of arrow worms hunting for zooplanktons. And even the amphipods that live at the deepest point of the ocean—in Mariana Trench—have been found with microplastics (the microfibers from synthetic clothing) in their bodies, according to the Plastic Soup Foundation.

Microplastics transfer from marine to terrestrial ecosystems by means of vital aquatic-terrestrial links. Terrestrial consumers, especially birds, and amphibians feed on the microplastic-contaminated consumers in aquatic systems like fish, and then they become a key source for continuing the food chain in terrestrial ecosystems.

The trophic transfer of microplastics, as intensively reviewed by Batel et al. (2016) and Mateos-Cárdenas et al. (2019), is critical for the ecological repercussions of the burgeoning microplastic pollution and is an emerging concern for human health. Experiments conducted by Athey et al. (2020) demonstrated that trophic transfer is a significant route of microplastic exposure exerting harmful influences in sensitive life stages. However, despite the ingested microplastic presence recorded in many important estuarine species (Alves et al. 2016; Bessa et al. 2018), the trophic microplastic transfer has not so far been investigated in estuarine ecosystems (Athey et al. 2020).

Seafood, comprising a variety of organisms, especially several fish varieties, consumed by large human populations worldwide is a major route for microplastics to transfer back to land. This direct anthropogenic transfer is of a more serious nature as it directly impacts human health.

Microplastics also enter into terrestrial systems in bulk through ocean waves sweeping the lands. Ocean sprays may also be possible carriers of microplastic pollution from aquatic to terrestrial systems. In the terrestrial environment, the microplastics meet the same fate of passing on from lower to higher trophic levels through grazing as well as detritus food chains (see Figure 3.2).

3.5   NANOPLASTICS IN THE OCEANS

Microplastics are not the final "product" of the plastic wastes in the ocean environment. These are further likely to be reduced into smaller particles up to less than 1 μm in size, known as nanoplastics. The nanoplastics originate as a result of fragmentation, photooxidation, and, possibly, biodegradations of the microplastics. Additionally, primary nanoplastics are perhaps emitted into the environment, too. From a chemical composition point of view, nanoplastics may comprise "virgin" polymers (for example, polyethylene), but can also comprise additives like the parent plastics, or can contain functional groups (for example, carbonyl moieties) added to the primary polymers in the processes of microplastics degradation (Wayman and Niemann 2021).

Presence of nanoplastics in the marine environment was discovered for the first time in 2017 (Halle et al. 2017). Nanoplastics detection in the environment and their quantification methodology is a recent investigation (Mintenig et al. 2018) and still seems to be flawed by some limitations in gaining an insight into environmental nanoplastic dynamics (Tushari and Senevirathna 2020). Widespread existence in the environment and potential adverse effects of nanoplastics on organisms (Shen et al. 2019) have recently drawn researchers' attention. Nanoplastics, according to the Plastic Soup Foundation, can enter the brains of the fish via the food chain, leading to abnormal behavior.

It is the sole reason why important information about the concentrations and distribution of nanoplastic particles in the ocean environments is not yet well understood. All the conventional sampling techniques used in the plastic particles up to microplastic sizes do not apply to the analyses of the nanoplastics. In order to understand the necessary environmental nanoplastic dynamics, appropriate analytic methods helpful in characterizing nanoplastics at the environmental level need to be developed and combined with suitable sampling strategies (Wayman and Niemann 2021).

Since there are no natural analogues for the nanoplastics, it is difficult to find out their transport, reactivity, toxicity, and definite fate in environmental systems (Besseling et al. 2019; Kögel et al. 2020; Wayman and Niemann 2021). Nanoplastics may undergo photooxidation (Wagner and Reemtsma 2019). Increased surface-to-volume ratio could make microbial degradation another possible fate of the nanoplastics

(Mattsson et al. 2015, 2018), which, however, has not yet been established (Wayman and Niemann 2021). Microbial degradation of the molecules released upon the photooxidation of the plastic particles has been demonstrated earlier; yet their identity has not been revealed (Romera-Castillo et al. 2018; Zhu et al. 2020).

The physical, mechanical, and chemical properties the nanoplastics possess are markedly different from the larger plastic particles. Owing to these properties, the dispersal patterns in the case of nanoplastics are different in comparison to the macro- and microplastics emanating from the same polymer. Plastic particles larger than or equal to µm are held in the control of buoyancy determined by density and shape. As a result, macro- and microplastics float, sink, or—in rare cases—are neutrally buoyant.

Dispersion of nanoparticles, however, is predominantly controlled by the nanoplastic collisions with water molecules and Brownian motion rather than governed by buoyancy properties (Hassan et al. 2015). The resultant colloidal behavior of nanoplastics causes them to disperse throughout the marine environment—the water column as well as in sediment. As a consequence, the nanoplastics would exert impact on ocean ecological niches as well as on the whole spectrum of ocean life. Nanoplastics can also undergo the processes of homoaggregation and heteroaggregation, that is, nanoplastics aggregation with the particles of same and different types of particles, respectively (Mattsson et al. 2015; Alimi et al. 2018; Oriekhova and Stoll 2018; Wagner and Reemtsma 2019).

Among these two types of aggregations, the heteroaggregation is likely to occur more frequently due to the ubiquity of other colloids in the water column (Alimi et al. 2018). The resulted heteroaggregates are pore-bearing, spongy, or mesh-like structures in which nanoplastics may be surrounded by inorganic and organic molecules taking a corona-like shape. However, this kind of nanoplastic aggregation is determined by the conditions such as surface charge modification, electrostatic interactions with ions and organic material in the surrounding aqueous medium, and bridging process (Cai et al. 2018; Singh et al. 2019). One of the consequences of the nanoplastic aggregation is reduced surface-to-volume ratio of the parent nanoplastic particles. Such a state of the microplastic pollution eventually impacts the surface chemistry of nanoplastics, generating complexities in the ecological processes in the marine environment.

3.6   MAJOR GAPS IN THE CURRENT KNOWLEDGE

Studies on the fragmentation of mega-, macro-, and mesoplastics into microplastics have so far left a gap of information about how the fragmentation leading to various sizes, up to microplastics and nanoplastics, occurs on the basis of the plastic shapes, such as spherical, elliptical, rectangular, ropes, haphazard, etc. Various operational factors in the marine environment, especially mechanical/physical factors, would be influenced by plastic shapes and so would the fragmentation of the plastics.

A large research gap exists in the quantitative analysis of different exposure routes of microplastics and their toxic effects on organisms (Wang et al. 2021). While pathways of microplastic fates in marine environments have been thoroughly sketched, other environmental components, especially atmosphere and soil, need to be investigated. Microplastic toxicity mechanisms also need to be adequately addressed through investigations. Large research gaps in the ecological risks associated with microplastics' relation to various ecosystems/environmental components—for example, open oceans, estuaries, sediments, soils etc.—should be bridged for a sound understanding of ecosystem interconnectivity and various dimensions of the implications of microplastic emissions.

Current techniques/methodologies are not capable of appropriately detecting nanoplastics in environmental samples. Microplastics originating from larger plastic sizes are not in the final stage of their fate before they get broken down constituent components, but the nanoplastics are. Associated ecological risks, bioavailability, and toxic effects of nanoplastics are distinguishable from those of the microplastics. Nanoplastics are potentially more hazardous than microplastics. The fate cycle of the nanoplastics is also likely to vary from that of the microplastics. Such knowledge gaps need be bridged by research studies with a focus on nanoplastics.

3.7 PERSPECTIVES FOR FUTURE STUDIES

Global demand for plastic items will continue unabated in the future. Plastic usage in several other fields, far more than at present, will also be witnessed in the future. Plastic production at an increased pace, therefore, is a likely future scenario. Increasing population pressures in coastal zones will escalate the rates of plastic emissions in the oceans. If sound eco-friendly plastic waste management systems involving renewable and environmentally safe alternatives to conventional plastics are not evolved and brought into public use, more of the plastic problems will precipitate and numerous more plastic pollution-related issues will come to the fore, and the future scenario will certainly be extremely dismal.

One of the future scenarios, albeit hypothetical, is that new strains of microbes, bacteria and fungi, would greatly help get rid of the problem. Biodegradation processes, however, are slower in temperate regions, a fact that portrays another scenario: microplastics' persistence in the oceans in temperate regions remains a serious problem.

Micro- and nanoplastics are likely to stay in the oceans due to continuous flow of plastic wastes, and consequent debris accumulation rates are likely to exceed that of biodegradation rates, despite the use of innovations in the plastic degradation processes involving well-sought physical, chemical, and biological factors.

Macro- and microplastics have been widely studied and their negative consequences to environment, ecology, and human health are well documented. Various physical, environmental, and biological factors contribute to the fragmentation and degradation of macroplastics into microplastics. But, that is not the end stage of the plastic fate. Microplastics are further liable to be fragmented into nanoplastics, especially owing to photooxidation and microbial actions on microplastics. The most probable future scenario with conventional plastic products in intensive use would be a spurt in the influx of nanoplastics, turning the oceans into a sort of "nanoplastics soup". Nanoplastics' impacts on organisms are more serious than those of microplastics. These are potentially more bioavailable than the microplastics (Wayman and Niemann 2021), spelling out a high degree of toxicity that can also directly enter into cells where they may cause deleterious effects leading to multiple health problems and disorders, including neurological ones, and changes in organisms' behavior. But there is also an increased probability nanoplastics will undergo degradation at a faster rate due to a higher surface-to-volume ratio. Nanoplastics are also subjected to aggregation mechanisms leading to their reduced stability in the marine environment. Future research projects would be aimed at establishing a clear understanding of microplastics and nanoplastics dynamics vis-à-vis their ecological risks and fates.

3.8 SUMMARY

Wastes generated from both primary and secondary plastic sources are accumulating in ocean waters due to various anthropogenic activities that fragment plastics into smaller particles ranging in size from macro- to micro- to nanoplastics and interacting with metals and organic pollutants. Microplastic pollutants tend to distribute in every sub-zone/layer of the marine ecosystems. Changes in the physical environment as a result of microplastics' presence induce alterations in the ecological processes, making them very unfavorable for marine organisms. The higher the microplastics concentrations, the greater the magnitude of microplastic-induced changes. The biology of aquatic environments corresponds broadly to variations in physical factors such as light, temperature, and water movements and to chemical factors, such as salinity and oxygen that would be influenced in many ways due to the presence of microplastics.

Microplastics frequently enter the human body through inhalation and ingestion of water and foods. Major toxicity effects are due to leakage into the respiratory and digestive tracts of additives and chemicals like plasticizers, light and heat stabilizers, antioxidants, flame retardants etc. added during the plastic production processes, and due to microplastic-associated contaminants, such as persistent organic pollutants (POPs), heavy metals, etc. Their toxic effects on the human body can result in, for instance, obesity, diabetes, heart attacks, infertility, sex malformation, cancer, behavioral changes, neurological disorders, etc.

Accidental ingestion of microplastics by organisms is common. Trophic transfer of microplastics beginning from autotrophs to top carnivores takes place in the microplastic-polluted environments. Microplastic transfer to terrestrial ecosystems takes place by means of several land-based consumers, such as birds and amphibians. Microplastic-contaminated seafood consumed by large human populations worldwide is a major route for the microplastics transfer to land. These pollutants could also enter into terrestrial systems in bulk through ocean waves and sea sprays.

Nanoplastics originate as a result of fragmentation, photooxidation, and possibly by microplastic biodegradation. Concentrations and distribution of nanoplastics in the ocean environments are not yet well understood. Increased surface-to-volume ratio could make microbial degradation another possible fate of the nanoplastics, which, however, has not yet been well established. Physical, mechanical, and chemical properties the nanoplastics possess are strikingly different from those of the larger plastic particles. The nanoplastics exert impact on ocean ecological niches as well as on the whole spectrum of ocean life.

Studies focused on plastic fragmentation based on heir shape, their comparative ecological impacts on various ecosystems/environmental components, and on microplastic toxicity mechanisms would help narrow down the research gaps. Our understanding of nanoplastics dynamics in the marine environment is very limited, which is attributable to lack of methodologies available to studying various aspects of their behavior and fates in the marine environments. The ecological impacts and toxicity effects of nanoplastics are more serious than those of the microplastics. It is necessary to resolve all these issues through investigations.

REFERENCES

Alimi, O. S., Budarz, J. F., Hernandez, L. M. and Tufenkji, N. 2018. Microplastics in aquatic environments: Aggregation, deposition, and enhanced contaminant transport. *Environmental Science and Technology*, 52:1704–1724. doi: 10.1021/acs.est.7b05559.

Alves, V. E. N., Patricio, J., Dolbeth, M., Pessanha, A., Palma, A. R. T., Dantas, E. W. and Vendel, A. L. 2016. Do different degrees of human activity affect the diet of Brazilian silverside *Atherinella brasiliensis*? *Journal of Fish Biology*, 89:1239–1257. doi: 10.1111/jfb.13023.

Andrady, A. L. 2011. Microplastics in the marine environment. *Marine Pollution Bulletin*, 62(8):1596–1605. doi: 10.1016/j.marpolbul.2011.05.030.

Athey, S. N., Albotra, S. D., Gordon, C. A., Monteleone, B., Seaton, P., Andrady, A. L., Taylor, A. R. and Brander, S. M. 2020. Trophic transfer of microplastics in an estuarine food chain and the effects of a sorbed legacy pollutant. *Limnology and Oceanography Letters*, 5(1):154–162. doi: 10.1002/lol2.10130.

Barrett, T. 2019. Microplastic pollution 'number one' threat to humankind. *Environment Journal*. https://environmentjournal.online/articles/microplastic-pollution-number-one-threat-to-humankind/

Batel, A., Linti, F., Scherer, M., Erdinger, L. and Braunbeck. 2016. Transfer of benzo[a]pyrene from microplastics to *Artemia nauplii* and further to zebrafish via a trophic food web experiment: CYP1A induction as visual tracking of persistent organic pollutants. *Environmental Toxicology and Chemistry*, 35:1656–1666. doi: 10.1002/etc.3361.

Bessa, F., Barria, P., Neto, J. M., Frias, J. P. G. L., Otero, V., Sobral, P. and Marques, J. C. 2018. Occurrence of microplastics in commercial fish from natural estuarine environment. *Marine Pollution Bulletin*, 128:575–584. doi: 10.1016/j.marpolbul.2018.01.044.

Besseling, E., Redondo-Hasselerharm, Foekema, E. M. and Koelmans, A. A. 2019. Quantifying ecological risks of aquatic micro- and nanoplastic. *Critical Reviews in Environmental Science and Technology*, 49(1):32–80. doi: 10.1080/10643389.2018.1531688.

Cai, L., Hu, L., Shi, H., Ye, J., Zhang, Y. and Kim, H. 2018. Effects of inorganic ions and natural organic matter on the aggregation of nanoplastics. *Chemosphere*, 197:142–151. doi: 10.1016/j.chemosphere.2018.01.052.

Eriksen, M, Libreton, L. C. M., Carson, H. S., Thiel, M., Moore, C. J., Borerro, J. C., Galgani, F., Ryan, P. G. and Reisser, J. 2014. Plastic pollution in the world's oceans: More than 5 trillion plastic pieces weighing over 250,000 tons afloat at sea. *PLoS One*, 9(12): e111913. doi: 10.1371/journal.pone.0111913.

Guo, X. and Wang, J. L. 2019. The chemical behaviors of microplastics in marine environment: A review. *Marine Pollution Bulletin*, 142:1–14. doi: 10.1117/1.OE.58.9.092608.

Halle, A. T., Jeanneau, L., Martignac, M., Jarde, E., Pedrono, B., Brach, L. and Gigault, J. 2017. Nanoplastic in the North Atlantic Subtropical Gyre. *Environmental Science and Technology*, 51:13689–13697. doi: 10.1021/acs.est.7b03667.

Hassan, P. A., Rana, S. and Verma, G. 2015. Making sense of Brownian motion: Colloid characterization by dynamic light scattering. *Langmuir*, 31:3–12.

Kannan, K. and Vimalkumar, K. 2021. A review of human exposure to microplastics and insights into microplastics as obesogens. *Frontier of Endocrinology*, 12:724989. doi: 10.3389/fendo.2021.724989.

Kögel, T., Bjorøy, Ø., Toto, B., Bienfait, A. M. and Sanden, M. 2020. Micro- and nanoplastic toxicity on aquatic life: Determining factors. *Science of the Total Environment*, 709:136050. doi: 10.1016/j.scitotenv.2019.136050.

Mateos-Cárdenas, A., Scott, D. T., Gulzara, S., van Pelt Frank, N. A. M., O'Halloran, J. and Jansen, M. A. K. 2019. Polyethylene microplastics adhere to Lemna minor, yet have no effects on plant growth or feeding by *Gammarus duebeni*. *Science of the Total Environment*, 689:413–421. doi: 10.1016/J.SCITOTENV.2019.06.359.

Mattsson, K. Hansson, L.-A. and Cedervall. 2015. Nano-plastics in the aquatic environment. *Environment Science: Processes Impacts*, 15:1712–1721.

Mattsson, K., Jocic, S., Doverbratt, I. and Hansson, L.-A. 2018. Nanoplastics in aquatic environment. In: Zeng, Y. (ed.) *Microplastic Contamination in Aquatic Environments*. London: Elsevier. 379–399.

Mintenig, S. M., Bauerlein, P. S., Koelmans, A. A., Dekker, S. C. and van Wezel, P. 2018. Closing the gap between small and smaller: Towards a framework to analyse nano-and microplastics in aqueous environmental samples. *Environmental Science: Nano*, 5:1640–1649.

Molles, M. C. 2005. *Ecology: Concepts and Applications*. Boston: McGraw Hill. 622pp.

Oriekhova, O. and Stoll, S. 2018. Heteroaggregation of nanoplastic particles in the presence of inorganic colloids and natural organic matter. *Environmental Science: Nano*, 5:792–799. doi: 10.1039/C7EN01119A.

Romera-Castillo, C., Pinto, M., Langer, T. M., Alvarez-Salgado and Herndl, G. J. 2018. Dissolved organic carbon leaching from plastics stimulates microbial activity in the ocean. *Nature Communication*, 9:1430. doi: 10.1038/s41467-018-03798-5.

SAPEA (Science Advice for Policy by European Academies. 2019. *A Scientific Perspective on Microplastics in Nature and Society*. Berlin: SAPEA. 173pp.

Shen, M., Zhang, Y., Zhu, Y., Song, B., Zeng, G., Hu, D., Wen, X. and Ren, X. 2019. Recent advances in toxicological research of nanoplastics in the environment: A review. *Environmental Pollution*, 252:511–521. doi: 10.1016/j.envpol.2019.05.102.

Singh, N., Tiwari, E., Khandelwal, N. and Darbha, G. K. 2019. Understanding the stability of nanoplastics in aqueous environments: Effect of ionic strength, temperature, dissolved organic matter, clay, and heavy metals. *Environmental Science: Nano*, 6:2968–2976. doi: 10.1039/C9EN00557A.

Singh, V. 2019. *Fertilizing the Universe: A New Chapter of Unfolding Evolution*. London: Cambridge Scholars Publishing. 285pp.

Singh, V. 2020. *Environmental Plant Physiology: Botanical Strategies for a Climate Smart Planet*. Boca Raton: CRC Press (Taylor and Francis). 216pp.

Tushari, G. G. N. and Senevirathna, J. D. M. 2020. Plastic pollution in the marine environment. *Heliyon*, 6(8): e04709. doi: 10.1016/j.heliyon.2020.e04709.

Wagner, S. and Reemtsma, T. 2019. Things we know and don't know about nanoplastic in the environment. *Nature Nanotechnology*, 14:300–301. doi: 10.1038/s41565-019-0424-z.

Wang, C., Zhao, J. and Xing, B. 2021. Environmental source, fate, and toxicity of microplastics. *Journal of Hazardous Materials*, 407:124357. doi: 10.1016/j.jhazmat.2020.124357.

Wang, J., Zheng, L. and Li, J. 2018. A critical review on the sources and instruments of marine microplastics and prospects on the relevant management in China. *Waste Management and Research*, 36(10):898–911. doi: 10.1177/0734242X18793504.

Wayman, C. and Niemann, H. 2021. The fate of plastic in the ocean environment—A minireview. *Environmental Science: Processes and Impacts*, 23:198–212. doi: 10.1039/d0em00446d.

Zhu, L., Zhao, S., Bittar, T. B., Stubbins, A. and Li, D. 2020. Photochemical dissolution of buoyant microplastics to dissolved organic carbon: Rates and microbial impacts. *Journal of Hazardous Materials*, 383:121065. doi: 10.1016/j.jhazmat.2019.121065.

4 Human Health Impacts of Microplastics Pollution

The extent of the rapidly spreading plastic garbage giving origin to microplastic pollution in the ocean ecosystems can be imagined from the fact that it has formed a sort of island in the Pacific Ocean. The Great Pacific Garbage Patch (GPGP), also familiar as the Pacific Trash Vortex and often referred to as the seventh continent, is a 79,000 metric tons of floating plastic waste forming an island in the central North Pacific Ocean (120°W–160°W, 20°N–45°N) covering an area of about 1.6 million km² (>3 times larger than the geographical area of Spain) with microplastic particles constituting 8% of the total garbage mass and 94% of the numbers of pieces out of about 1.8 trillion floating pieces (Lebreton 2018). The fact is that the plastic debris and the originating microplastic pollution are not there only in the vortex, but everywhere in the ocean ecosystems. Microplastics' abundance and their worldwide distribution appear to have fixed our world into a historical epoch many scientists define as "Plasticene" (Reed 2015; Campanale 2020). This extent in the spurt of microplastics in the planet's largest environmental component itself narrates the story that humans cannot escape from the exposure and effects of this pollution.

Microplastics (plastic particles <5 mm in size) comprise only 3% of the environmental pollution. However, the microplastics pollution has potentially negative implications on ecosystems, especially on aquatic ecosystems and eventually on human health. This pollution, like other kinds of pollution—air, water, and soil—has no role of natural factors, but is purely of anthropogenic origin. The occurrence of the microplastics pollution has, in fact, emerged from the manufacture of plastic materials for the sake of convenience. Impacts of microplastics on human health are serious in nature and treatment of the affected people is also quite complex.

4.1 HUMAN EXPOSURE TO MICROPLASTICS POLLUTION

The microplastics emanating from human activities are primary microplastics, whereas those originating from the fragmentation of large-size plastics are secondary microplastics (European Food Safety Authority 2016). Our markets, commercial establishments, industries, educational institutions, research and development centers, public places, and households inevitably depend on plastic items. A variety of domestic gadgets, furniture, utensils, etc. are made up of plastics. Exposure of most of the planet's life, not just human species, to microplastics pollution becomes almost inevitable looking at the scenario of overwhelming microplastic presence in all the Earth's ecosystems including where life exists with an extreme degree of biodiversity and where life hardly exists.

DOI: 10.1201/9781003312086-4

Microplastics are prevailing everywhere and in everything; for example, surface waters, saline waters, running waters, standing waters, bottled waters, wastewaters, sediments, soils, the Arctic sea, Antarctic ice, indoor air, the atmosphere, salt, sugar, honey, beverages, and other food products. Microplastic pollution is being realized on global scale and is a significant environmental and public health concern (Zhang et al. 2017; Teng et al. 2019; Kannan and Vimalkumar 2021).

Huge amounts of plastics reach our kitchens along with food items and processed and packaged foods. In households, plastics are also directly available in the form of microplastics granules through frequently used cosmetics, scrubbers, etc. (Sharma and Chatterjee 2017) and thus become a source for direct exposure of human beings (Figure 4.1). The Covid-19 pandemic has caused more of the polymers to be used to combat the pandemic, such as disposable face masks, with potential environmental and human health risks (Du et al. 2022), and personal protective clothing (Uddin et al. 2021).

There are two ways humans get ubiquitously exposed to microplastic pollution, namely inhalation of contaminated air and dust, and ingestion of contaminated water and foods. These pathways through which microplastics make entry into humans are well established. Human beings are bound to ingest millions of microplastics annually, or several milligrams daily (Kannan and Vimalkumar 2021). Medical devices in hospitals and feeding through bottles expose newborns to microplastics ingestion.

Microplastic particles ubiquitously present in the food web (Cox et al. 2019; Senathirajah et al. 2021) are, undoubtedly, the gravest dimension of the microplastic pollution. It is due to this aspect of the ecosystems that the microplastics are steadily making access to perhaps all the species thriving in the ecosphere. Microplastic invasion of humans takes place through various pathways, linked and non-linked with the food web. Varying amounts of microplastics ingested by humans have been studied by various workers (see Table 4.1).

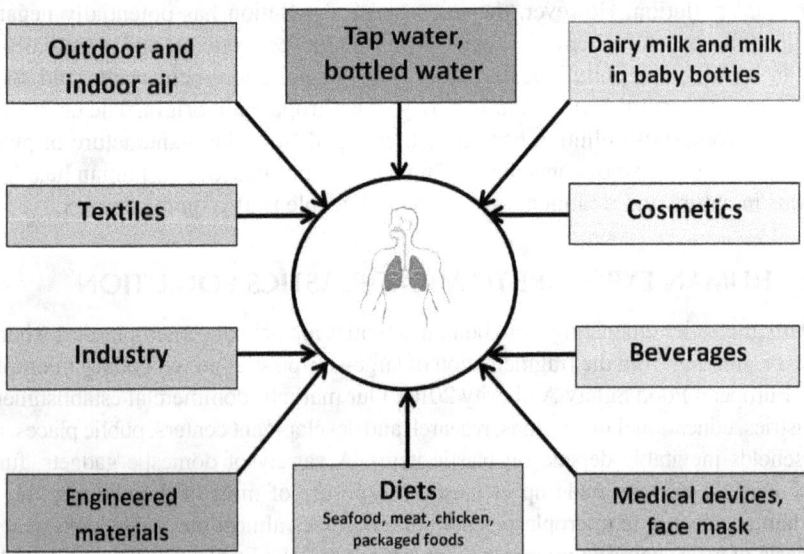

FIGURE 4.1 Human exposure to various sources of microplastics.

TABLE 4.1

Microplastics in various foods and food supplements

Sources of microplastic pollution	Unit	Values	References
Bivalves (e.g., mussels, clams, scallops, etc.)	MPs per gram	0.2–4	Zarus et al. (2020); Fournier et al. (2021)
Shrimp/Crab	MPs per gram	0.75	Zarus et al. (2020); Fournier et al. (2021)
Fish	MPs per gram	1–7	Zarus et al. (2020); Fournier et al. (2021)
Seafood	MPs per gram	1.48	Cox et al. (2019)
Chicken (in gizzard)	MP particles	63	Lwanga et al. (2017)
Chicken (in crop)	MP particles	11	Lwanga et al. (2017)
Packaged meat (white chicken breast and turkey)	MPs per kg	4.0–18.7	Kedzierski et al. (2020)
Dairy milk	MPs per liter	3–11	Kutralam-Muniasamy (2020)
Milk in PP feeding bottles during sterilization	MPs released per liter	16,000,00	Li et al. (2020)
A single plastic teabag in hot water at 95°C	MPs released into a single teacup	11,600,000,000	Hernandez et al. (2019)
A single plastic teabag in hot water at 95°C	NPs released into a single teacup	3,100,000,000	Hernandez et al. (2019)
A single plastic teabag in hot water at 95°C	MPs released into a cup of tea	23,000,00	Cited by Kannan and Vimalkumar (2021)
A single plastic teabag in hot water at 95°C	NPs released into a cup of tea	14,700,000,000	Cited by Kannan and Vimalkumar (2021)
Honey	MPs per gram	0.10	Cox et al. (2019)
Honey	Plastic fibers per kg	40–660	Zarus et al. (2020); Fournier et al. (2021)
Sugar	MPs per gram	0.44	Cox et al. (2019)
Sugar	Plastic fibers per kg	217	Zarus et al. (2020); Fournier et al. (2021)
Salt	MPs per gram	0.11	Cox et al. (2019)
Salt	MPs per kg	0–19,800	Zarus et al. (2020); Fournier et al. (2021)
Single use of facial scrub, a cosmetic	Microbeads (MPs)	4594–94,500	Napper et al. (2015)
Synthetic textile	Microfibers released in a 6-kg wash load of acrylic fiber	700,000	Xu et al. (2020)
Polyester fabric	Microfibers on first washing cycle	210,000–13,000,000	Sillanpää and Sainio (2017)
Polyester washing by household washing machine	Annual emission of microfibers, number (weight)	1.0×10^{14} (154,000 kg)	Sillanpää and Sainio (2017)

Note: MP—microplastic; NP—nanoplastic; PP—polypropylene.

A food chain human beings depend upon is linked with several different inter-mediates, namely foods processing, chemical treatment, and distribution. There are chances that until the food reaches consumers, it could be potentially contaminated by microplastics.

Textile fibers with 33% microplastics were found in concentrations ranging from 1.0 to 60.0 fibers per cubic meter and 0.3 to 1.5 fibers per cubic meter in atmospheric fallout from indoor air and outdoor air, respectively (Dris et al. 2017). In an environ-ment infested with microplastics, humans would unintentionally consume microplas-tic particles even if their food is not contaminated by them.

Senathirajah et al. (2021) estimated global average of microplastic ingestion by humans through various exposure pathways at 0.1–5 g per week (i.e., about 5–260 g per annum). In her study on airborne plastics and their consequences for human health, Prata (2018) detected microplastics in atmospheric fallout that are eventually consumed by human individuals through inhalation and dust-laden particles. A daily per capita microplastic intake rate equal to 184 and 583 ng for children and adults respectively through nine varying exposure sources were predicted by Nor et al. (2021) in their probabilistic lifetime exposure model.

Humans ingest or inhale around 50,000 microscopic plastic particles in a year (Kurtz and Sample 2021). A study aimed at enumerating microplastics consumed by humans conducted by Cox et al. (2019) estimated per capita annual microplastics consumption based on age and sex between 39,000 and 52,000 particles. The number of the particles increased to the range between 74,000 and 121,000 when inhala-tion of the particles was added to those consumed through foods. Drinking bottled water led to additional 90,000 particles in the list of human-consumed microplastics. Additional microplastic particle counts through drinking tap water were only 4000, that is, significantly less than those consumed through plastic bottles (Table 4.2).

Freshwater resources are also quite rich sources of microplastics and drinking water deriving from these sources poses risk of human exposure to the pollutants. Groundwater sources have been found to contain lower concentrations compared to tap and bottled water (Koelmans 2019). Single-use plastic products in use in Germany have been found containing polyethylene and polyethylene terephthalate particles of 5–20 μm size, and microplastic concentrations in drinking water were 118 and 14 particles in returnable and single-use bottles, respectively (Schymanski et al. 2018).

Indoor air appears to be the major source to pass on microplastic particles into the human body via inhalation (Cox et al. 2019; Wang et al. 2021). Atmospheric con-centration of microplastics is reported to be 9.8 microplastics per cubic meter (Cox et al. 2019). With a daily inhalation rate of 15 cubic meters, annual average of inha-lation exposure to microplastics would be equal to 53,700 microplastic particles per person (Kannan and Vimalkumar 2021). Such figures derived from different studies are presumed to be conservative. Airborne microplastics concentration, of course, would vary according to sampling methods as well as on account of environmental and anthropogenic factors and, therefore, these figures would also vary from place to place. However, it is clear that human exposure to airborne microplastic pollu-tion matters a lot. Main sources of the airborne microplastic particles in indoor as well as outdoor air include plastic degradation, building materials, synthetic textiles, waste incineration, landfills, etc. (Dris et al. 2015, 2016). Research found the major

TABLE 4.2

Annual per capita microplastic intake from various sources by a human individual.

Sources of microplastic pollution	Unit	Values	References
Food, water, dust, air	Numbers per annum	74,000–121,000	Cox et al. (2019)
Bottled water	Numbers per annum	90,000	Cox et al. (2019)
Sea salt	Numbers per annum	37–1000	Prata (2018)
Tap water	Numbers per annum	4000	Prata (2018)
Shellfish	Numbers per annum	11,000	Prata (2018)
Airborne microplastic particles through inhalation	Numbers per annum	53,700	Cox et al. (2019)
Indoor air inhalation	Numbers per day	26–130	Prata (2018)
Inhalation	Numbers per day	272	Vianello et al. (2019)
Ingestion or inhalation	Numbers of microscopic plastic particles per year	50,000	Kurtz and Sample (2021)
Drinking water	Numbers per day in male children, male adults, female children, and female adults, respectively	48, 55, 47, and 51	Cox et al. (2019)
Milk in PP bottles fed to babies up to 12 months old	Number per capita per day	14,600–4,550,000	Li et al. (2020)
Milk in PP baby feeding bottles	Number per day in milk-fed babies	3,000,000	Li et al. (2020)
All exposure pathways	Grams per week	0.1–5	Senathirajah et al. (2021)
Indoor dust inhalation	ng/kg-bw/day for PET in infants	150,000	Zhang et al. (2019)
Indoor dust inhalation	µg/kg-bw/day	6.5–8.97	Zhang et al. (2019); Wang et al. (2021)
Indoor dust ingestion	mg/day for a person weighing 70 kg	10	Kannan and Vimalkumar (2021)
Liquid soap with microbeads	PE microplastics, mg/capita/day	2.4	Gouin et al. (2011)
Facial scrub with microbeads	PE microplastics, mg/capita/day	0.5–215	Napper et al. (2015)

Note: bw—bodyweight; ng—nanogram; PE—polyethylene; PET—polyethylene terephthalate; PP—polypropylene.

airborne microplastic concentrations would also vary from place to place: for example, in Denmark, polyethylene (5–28%), polypropylene (0.4–10%), polyethylene terephthalate (59–92%), and nylon (0–13%) (Vianello et al. 2019); in China polyethylene terephthalate (PET) and acrylic fibers were mostly found (Zhang et al. 2020a). PET-based microplastics have been found to be ubiquitous. PET concentration has been recorded to be ranging from 38 to 120,000 µg g^{-1} and PET has also been revealed as the most abundant chemical in indoor dust (Zhang et al. 2019). The flux of microplastics in an urban area has been recorded two-fold higher compared to that in a suburban area (Dris et al. 2016). Further, the air concentration in the rainy season is reported to be higher than in the dry season (Dris et al 2015).

Human exposure to indoor dust ingestion has also been attempted to be measured in terms of bodyweight (bw). Geometric mean exposure doses to microplastics via indoor dust ingestion, thus, have been recorded to be in the thousands of nanograms (ng) per kg-bw per day (Kannan and Vimalkumar (2021). The highest figure for infants' exposure to polyethylene terephthalate obtained in this regard was 150,000 ng per kg-bw (Zhang et al. 2019), a value that corresponds to approximately 10 mg per day for a person weighing 70 kg (Kannan and Vimalkumar 2021). Another study reported microplastic doses through inhalation in the contaminated air in the range of 6.5 to 8.97 µg per kg-bw per day and were found to be 3- to 50-fold higher in infants and toddlers than in adults (Zhang et al. 2019; Wang et al. 2021). In yet another study, average per capita per day inhalation exposure has been estimated in the range between 26 and 227 microplastic particles (Prata 2018) which, on an annual basis, would come out to be between 9490 and 47,450 microplastic particles.

Based on average water consumption patterns among male adults, female adults, male children, and female children, Cox et al. (2019) in their study elicited mean daily microplastic intake and the values that came to the fore were 55, 51, 48, and 47 microplastics, respectively. The microplastic dose entered into the human body through drinking bottled water was observed to be 22 times higher than through tap water.

A variety of market-based drinks and beverages, such as wine, beer, bottled tea, energy drinks, etc., are also a rich source of microplastics. Almost all people in our times are fond of such drinks as per their choice and, therefore, microplastic consumption through them also becomes inevitable. The fact that humans get exposed to microplastics by being dependent on these beverages has been revealed through a study by Shruti et al. (2020). According to this study, microplastic concentrations in the beer samples from the USA, Mexico, and Germany were found to be in the range of 0–14, 0–28, and 10–256 particles per liter, respectively. The study also revealed that wine from Italy had 2563–5857 particles per liter. Energy drinks, soft drinks, and tea from Mexico were found with 11–40 microplastic particles per liter. Human populations throughout the world are fond of tea with some societies being more addicted to this beverage than others. The extent of human exposure can be imagined from a research report that states that a single plastic teabag dipped in a cup of hot water at 95°C releases about 11.6 billion microplastics and 3.1 billion nanoplastics (Hernandez et al. 2019) (see Table 4.1).

After inhalation through contaminated air and ingestion through drinking water, food is the key source for humans' exposure to microplastic contamination (Kannan and Vimalkumar 2021). Foods brought home from the fields may be safe, but risks of exposure to macroplastics increase when processed foods, fast food, and packaged foodstuffs bought from the market are consumed at home.

There is a very long list of cosmetics widely used across the world that contain microbeads, the microplastic particles generally less than 1 mm in size used as abrasives. A few examples are hair bleaches, hair colorants, deodorants, nail polishes, lipsticks, sunscreens, shower gels, scrubs, soaps, shampoos, skin creams, etc. (Cheung and Fok 2016; Kannan and Vimalkumar 2021). Microbeads have been discovered in abundance in coastal waters, such as in Hong Kong (So et al. 2018), which is indicative of the microplastics pollution spreading from excessive use of cosmetics. Microbeads amounting to 4130 tons were used annually in soaps in the European Union, as documented by Kannan and Vimalkumar (2021). In cosmetics, the microbeads are mixed in varying proportions, ranging from 0.5 to 12%, and about 93% of the microbeads in the cosmetics comprise polyethylene while some may contain PET, PET and PMMA, or nylon.

Due to the addition of microbeads in liquid soaps, people are readily exposed to microplastics. Daily per capita microplastic consumption rates due to use of liquid soaps in USA were estimated at 2.4 mg polyethylene particles (Gouin et al. 2011). A facial scrub used once leads to the release of as many as 4594 to 94,500 microbeads (microplastics) into the water medium (Napper et al. 2015). Polyethylene microbeads in the facial scrubs in the United Kingdom (1–10 g per 100 ml) are consumed by users at the rate of 0.5 to 215 mg per person per day (Napper et al. 2015).

Synthetic textiles release huge numbers of microfibers when they are washed. A 6-kg of wash load of acrylic fiber has been observed releasing as many as 700,000 microfibers (Xu et al. 2020) and 210,000 to 13,000,000 microfibers upon washing polyester fabric in the first cycle (Sillanpää and Sainio 2017). The annual emission of polyester from household washing machines was estimated at 1.0×10^{14} (154,000 kg) in Finland (Sillanpää and Sainio 2017). The Covid-19 pandemic has imposed use of face masks as an essential precautionary measure. Use of plastics in the face masks may have added an extra load of microplastics through inhalation. However, this is yet to be verified through research (see Tables 4.1 and 4.2).

4.2 MICROPLASTICS' PRESENCE IN SEAFOOD

Seafood is most likely to be contaminated *in situ* as the oceans and seas are the ultimate dumping grounds of the plastics. The microplastic-contaminated seafood has been extensively studied and microplastic concentrations have been reported as high as 200,000 particles per kg of the food (Elizalde-Velázquez and Gómez-Oliván 2021). An analysis of table salt samples from various countries revealed that the table salt from Asian countries recorded higher microplastic concentrations compared to the samples from the countries outside the Asian continent (Kim et al. 2018).

Seaweed, fish, bivalves, crustaceans, chickens, salt sugar, and honey all have been found containing microplastics (Liebezeit and Liebezeit 2013; Mathalon and Hill 2014; Neves et al 2015; Li et al. 2016; Karami et al. 2017; Bessa et al. 2018; Karbalaei 2018; Naji et al. 2018; Oßmann et al. 2018). Foods packed in plastic wraps or plastic containers, such as is the case of fast foods, are invariably contaminated by microplastics. Commonly found microplastics in the packaged food items, as documented by Kannan and Vimalkumar (2021), emanate from polyethylene, polyethylene terephthalate, polyurethane, polystyrene, polyvinyl chloride, polypropylene, polyamide, polymethyl methacrylate, and styrene acrylate (see also Tables 4.1 and 4.2).

Potential global exposure of infants up to 12 months old to microplastics can be imagined looking into a report by Li et al. (2020) based on the survey of 48 regions that reveals that per capita per day microplastic consumption rates in infants range between 14,600–4,550,000 particles. Sulfone polymers used in the dairy industry for filtration purpose are the major sources of microplastics contaminating the milk (Kutralam-Muniasamy 2020; Kannan and Vimalkumar 2021).

Microplastics may contain nearly 4% additives, and plastic waste may also be adsorbing pollutants in their surrounding environment. In their nature, the contaminants as well as the additives may be organic and inorganic types. Microplastics present in seafood might impact human exposure to additives or contaminants (European Food Safety Authority 2016). Fish have been found containing high concentrations of microplastics in their stomach and intestines. Since the alimentary canal of fish is removed, they may not be a source for the exposure of consumers to microplastics. However, exposure risks are there for those who consume crustaceans (for example, shrimps, crabs, etc.) and bivalve mollusks (for example, oysters, mussels, clams, scallops, etc.) as the digestive tracts of these organisms are consumed.

Microplastic particles may attract accumulation of hazardous chemicals in high concentrations. Polycyclic aromatic hydrocarbons (PAHs) and polychlorinated biphenyls (PCBs) are examples. Bisphenol A (BPA), a compound used in food packaging, may leave its residues on the foods. Consumers incorporating the foods contaminated with chemical-laden microplastics would have these pollutants transferred to their tissues (European Food Safety Authority 2016).

The blue mussel (*Mytilus edulis*), the common mussel (a marine bivalve mollusk) cultivated for human consumption, may ingest microplastics of 2–10 µm size, according to a report of European Food Safety Authority (2016). This report also mentioned that average microplastic concentration per gram was up to 0.2–4 in bivalves, 0.75 in shrimp, and 1–7 in fish.

4.3   NANOPLASTICS' PRESENCE IN FOOD

Nanoplastics with size ranging from approximately 1 to 100 nm, or 0.001 to 0.1 µm, come into existence due to the fragmentation of microplastics, or they originate from engineered materials. Methods for the identification and quantification of microplastic particles in foods as well as in other materials have been evolving. There is a limited body of literature explaining the occurrence of the microplastic particles in foods. On the other hand, no methodology has been developed for understanding the presence and for quantification of nanoplastics in foods. No considerable occurrence data about nanoplastics in food are available. Their explanations are largely on a theoretical basis. Toxicity and toxicokinetics data about nanoplastics are also lacking (European Food Safety Authority 2016).

Mechanical fragmentation of low-density polyethylene (LDPE) produces an extremely large number of microplastics (10^4 to 10^6 items per gram of LDPE) and photochemical aging of LDPE nanoplastics produces an estimated 10^{10} nanoplastics per gram of LDPE from secondary microplastics (Sorasan et al. 2021).

More toxic than microplastics, nanoplastics have been detected in both polar regions of the planet and they are now pervasive around the world (Carrington 2022). Nanoparticles in Greenland arrived through blowing winds and in the Antarctica these have been transported by ocean currents. In Greenland, about half of the nanoparticles were found to be composed of polyethylene, about a quarter was tire dust, and about a fifth was polyethylene terephthalate. Polypropylene, polyvinyl chloride, and polystyrene are the other nanoplastics discovered in Greenland in small proportions. In Antarctica 55.2% of the nanoplastics comprised polyethylene, 28.4% polypropylene, and 16.4% polyethylene terephthalate. The average nanoplastics concentration reported was 46 ng ml^{-1} of melted surface snow, and deposition rate at the surface of snow was 42 kg km^{-2} year^{-1} (Materić et al. 2021).

Toussaint et al. (2019) reviewed several publications about micro- and nanoplastics present in animals belonging to more than 200 species that humans use to derive food from (sea fish, freshwater fish, mollusks, crustaceans, turtles, birds, and chickens) and food products connected with the human food chain (sea salt, sugar, honey, etc.) that could contribute to the uptake of micro- and nanoplastics, directly or indirectly.

Nanoplastics basically are not the original ingredients of food. They originate from non-food sources and reach the food aided by various mechanisms, such as through trophic links or during processing aids and distribution routes, or by means of machinery, equipment, and textiles. Air and water are the major environmental factors to carry the microplastics and nanoplastics to the organisms. What happens to nanoplastics when the seafood is processed? Data on this are not available. As the nanoparticles can enter a cell and become an inseparable part of it, the probability is that they remain intact during other processes needed before food consumption, such as cooking and baking.

What is the fate of nanoplastics in the gastrointestinal tract? It is not known precisely. Information on toxicokinetics gives an idea only about absorption and distribution, but there is no information about their metabolism and excretion (European Food Safety Authority 2016; Toussaint 2019). Nanoplastics originate from further fragmentation of microplastics aided by several factors, but whether and how the microplastics reduce to nanoplastics is not known. One such possibility is likely to occur in the gizzards of avian species where mechanical breakdown of food stuffs takes place, although no information is available in this regard. Some toxic effects of engineered nanomaterials have been brought to the fore (European Food Safety Authority 2016). However, how the nanoplastics cause toxic effects in the human body is not clearly understood. Nanoplastics can make entry into body cells. However, their consequences on human health are yet to be precisely explained. Since, upon entry into a cell, the nanoplastics become part and parcel of the basic functional unit of an organism, its intended functions would be affected to a certain extent. Therefore, it is most likely that the nanoplastics entering into the human body through contaminated foods are causing ill effects on human health, a fact that needs to be clarified by generating information based on research.

M cells, the characteristic epithelial cells lacking microvilli, transport the nanoplastics from the gut into the blood stream and from there into the liver and gallbladder via the lymphatic system (Bergmann et al. 2015). Two nanoplastic characteristics, size and hydrophobicity, help nanoplastics enter into the lungs and gastrointestinal tract through the placenta and blood-brain barrier (Seltenrich 2015). These organs are the potential sites for the nanoplastics to inflict their harmful effects. Large surface area to volume ratios cause the nanoplastics to assume a pretty high degree

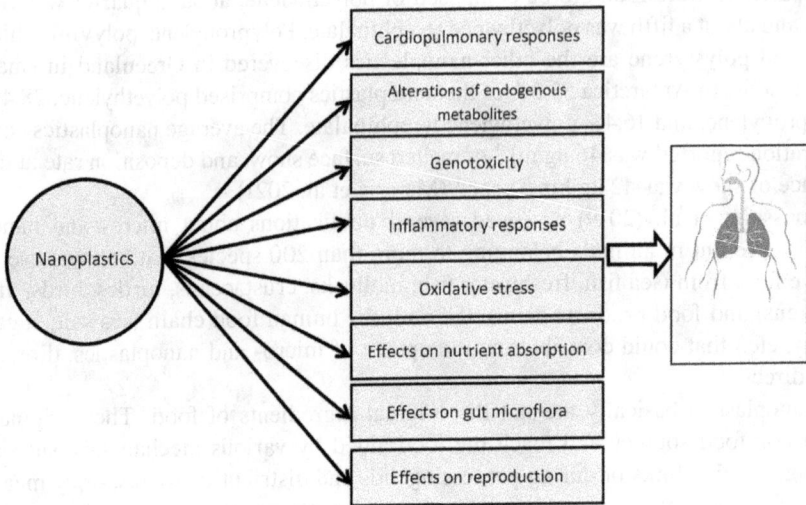

FIGURE 4.2 Nanoplastics effects on human health.

Source: Based on the information documented by Smith et al. (2018).

of chemical reactivity. Liver, lung, and brain cells are affected by nanoplastic toxicity, according to a review study by Smith et al. (2018). Nanoplastics can be the causal agents of inflammation and cellular barriers. These particles may also cross the blood-brain barrier and the placenta (Sharma 2020). The major effects of the systemic distribution of nanoplastics on human health are shown in Figure 4.2.

4.4   FOOD SECURITY

Food security is a critical global issue pertaining to human well-being. This issue, unquestionably, needs to be resolved at global level. The term "food security" is defined extensively by several workers from many disciplines as well as by social activists. For example, Smith, Pointing, and Maxwell identified about 200 different definitions of food security, and Gentilini 205 (Simon 2012). The most commonly accepted definition, however, is the one approved by the World Food Summit (WFS) (FAO 1996): "Food security exists when all people, at all times, have physical, social and economic access to sufficient, safe and nutritious food which meets their dietary needs and food preferences for an active and healthy life".

In the FAO's definition of food security, the word "safe" encompasses the quality of food; food needs be unspoiled by the presence of undesirable and unhealthy or toxic physical, chemical, and/or biological pollutants. In the context of microplastics presence in foods, such as in seafood, meat obtainable from a number of terrestrial animals, milk and milk products, and other food ingredients—directly or through trophic hierarchy—it can be inferred that the "safety" inherent with the food security definition is compromised. When microplastics pollution has emerged as ubiquitous in all the components of the environment, non-presence of primary and secondary microplastics in food or food ingredients during food processing, packaging, and distribution processes should be ascertained. However, avoidance of microplastic contamination of food through the food chain remains a challenge to be met.

The FAO identifies four pillars of food security: availability, access, utilization, and stability. In the concept of food security "utilization of food" addresses the body's ability to make the most out of the nutrients in consumed foods. Utilization of foods for the body's structural and functional needs through metabolic processes via digestion is affected by factors such as food processing prior to consumption; presence of undesirable, toxic, and non-food ingredients. Metabolism subject to "utilization of food" is influenced by the quality of food. Food quality deteriorated due to microplastics' presence is bound to affect food utilization because of two reasons:

1. Microplastics have no nutritive value and are no source of energy and thus their fate in the metabolic cycle is predictable—leading to adverse consequences.
2. Microplastics may harbor toxic chemicals and carry pathogens of various kinds which could impair normal metabolic processes.

Colonization of microplastics exerts population-level impacts (Wright et al. 2013). Plastic and microplastic pollution prevailing in aquatic and agro-ecosystems is threatening food security. Marketing of packaged ultra-processed and fast foods is generating ever-increasing plastic pollution. Microplastics entering into food chains is posing threat to all life, not only to human beings. The food system itself seems to be creating more and more plastic pollution. Food security-related processes from postharvest processing to distribution, in fact, are adding to microplastic pollution.

Annual environmental damage due to plastics is estimated at US Dollar 75 billion and food and beverage companies alone are responsible for 23% of the cost (Kurtz and Sample 2021). Environmental concerns emerging out of a situation of microplastic pollution call for analyzing our food security related issues with the ubiquitous problem of plastics and microplastics. "Stability", another component in the FAO's food security concept, underlines the need of food security in all times to come, that is, sustainable food security, which, emanating from the concept of "sustainable development" proposed by the World Commission on Environment and Development (WCED), or the Brundtland Commission, in its 1987 Report, *Our Common Future* (Singh 2019), means meeting food security of present generations without compromising the abilities of the future generations to fulfill their own food security needs.

A major problem the aquatic ecosystems and agro-ecosystems face is the overwhelming use of cheap and single-use plastics that are used only once and then thrown away. A major chunk of the single-use plastics, nearly 98%, comes from virgin fossil fuel-based feedstocks, especially fracked natural gas. A large proportion of the disposable plastics end up in the open environment or landfills. About 12% are burned and only 9% of them are recycled. Keeping pace with the current rate of growth, plastics will account for approximately 20% of fossil fuel demand by 2050, much higher compared to nearly 6% in 2016. One of the serious consequences of this state of things is that by 2050 its contribution to the greenhouse gas emissions budget will increase to about 15% (up from 1.7 Gt of CO_2e in 2015 to 6.5 Gt CO_2e) (Kurtz and Sample 2021).

The mess of plastics and microplastics in the marine environment leads to the reduction in secondary productivity, a state that impacts humans' seafood-dependent food security. Livestock regularly ingest plastic wastes. The indigestible plastic pieces consumed by ruminants "grazed" from the plastic waste heaps are detrimental to their health and cause adverse effects on their growth rate, draughtability, milk yields, and quality of the milk. Their morbidity and mortality rates are also increased. Adverse impacts of plastic pollution on livestock and livestock-dependent communities such as in poor countries include increases in food insecurity on the one hand and decreases in health due to microplastic particles' toxins and pathogens entering into human bodies on the other.

Microplastic pollution seems to hamper food security by contributing to breaking down ecological integrity. Ecological integrity is ensured by nature's dynamics expressed through vital flows of nutrients (biogeochemical cycles), water flows (hydrological cycles), gaseous flows, gene/pollen grain flows (pollination), and energy flows into autotrophs (photosynthesis and chemosynthesis) and through trophics (food chains) (Singh 2019, 2020). Microplastics pollution has been found to impact honeybees, the most crucial mediator of cross-pollination. A research study conducted by Al Naggar et al. (2021) in Ecuador found 12% of honey samples were contaminated with polyethylene (PE), polypropylene (PP), and polyacrylamide (PAM). Presence of microplastics has been detected in honeybees in the apiaries in Copenhagen and surrounding cities. The study elaborated that polystyrene (PS) microplastics reduced diversity of honeybee gut microbiota to follow changes in gene expression relating to detoxification, oxidative damages, and immunity. Such adverse impact of microplastics on honeybees and possibly on other vital pollinators is directly concerned with agricultural production and food security.

4.5 TOXICITY TO HUMANS

Single-use plastics debris items, notably bags, bottles, wrappers, food containers, and cutlery, seem to be like invasive species in water bodies. There is a piece of plastic waste per every meter of shoreline and as many as 18,000 floating pieces per km^2 of ocean (Kurtz and Sample 2021). Evidences indicate that humans constantly inhale and ingest microplastics originating from different sources in considerably high amounts (see Table 4.2). However, whether the microplastics and nanoplastics pose substantial risk to human health is not precisely understood (Dick Vethaak and Legler 2021). Most microplastics go in and out of most organisms. Laboratory experiments reveal that microplastics in high concentration and under specific conditions can induce physical and chemical toxicity (SAPEA 2019).

Experiments on many aquatic animals under laboratory conditions showed considerable negative effects of microplastics on food intake, growth rates, reproduction, and survivability. However, there is no evidence whether the animals with microplastic exposure in nature express similar effects (SAPEA 2019). Critchell and Hoogenboom (2018) through their laboratory experiments on planktivorous reef fish (*Acanthochromis polyacanthus*) concluded that growth and body conditions of the fish are negatively affected when their food is replaced by plastic particles and that the plastics become more of a problem when these break up into smaller and smaller particles. It is not microplastic alone, but the changes in the environment induced along with its presence—such as decreased phytoplankton concentration in marine ecosystems—that exert greater impact on the health of the organisms rather than the microplastics alone.

Serious health risks to humans in the communities living close to plastic manufacturing facilities or petrochemical plants are also evident and such risks disproportionately impact vulnerable populations (Kurtz and Sample 2021).

4.5.1 PHYSICAL EFFECTS OF MICROPLASTICS EXPOSURE

Millimetric and submillimetric sizes of microplastics facilitate their getting easily ingested by all the organisms exposed to the environment contaminated by these particles. Aquatic animals are especially susceptible to microplastic ingestion. Colonization of microplastics marks their effects at the population level, and size, shape, and abundance have influence on microplastic uptake (Wright 2013). The ubiquitous nature of microplastics makes them get inhaled and ingested by terrestrial animals—and eventually by humans—directly and through the food chain. Microplastics can also enter the body through dermal contact.

Ingestion of microplastics leads to microplastic toxicity which happens through three possible pathways, as also outlined in Figure 4.3, namely,

1. Ingestion stress.
2. Additive leakage.
3. Absorption of contaminants by microplastics.

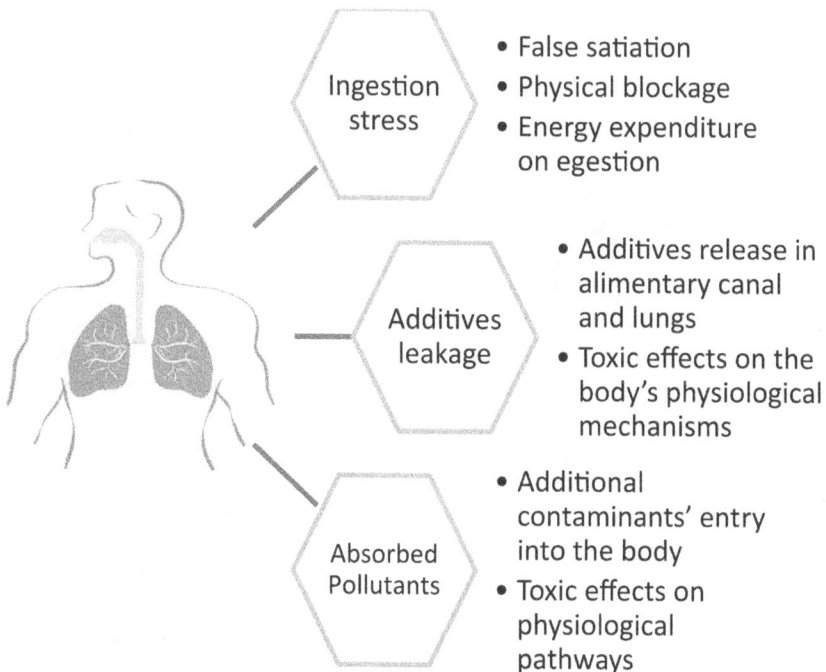

FIGURE 4.3 Pathways to toxicological effects of microplastics.

Microplastics ingestion causes stress in response to energy expenditure required for egestion, physical blockage, and a sort of false satiation. In contaminated diets these contribute to the bulk of food. As a result, an individual would feel satiation without consuming the required food amounts. This would also lead to deficiency of the required nutrients and energy level of an individual knowingly or unknowingly dependent on diets contaminated by microplastics. Physical blockage created by microplastics resulting in indigestion, constipation, coughing, etc. would add to the body stress. As microplastics are undesirable foreign elements making entry into the body, the digestive and excretory systems need to get rid of them, a process that needs expenditure of extra energy, thus putting an individual under constant stress.

4.5.2 EFFECTS OF CHEMICAL ADDITIVES

Effects of certain chemicals on human health by means of microplastics inhaled, ingested, and/or through dermal contacts are due to two fundamental characteristics pertaining to the microplastic particles, namely,

1. Microplastics have additives not bound with the constituent polymers. These additives get released in our digestive and respiratory systems. Leakage of additives creates a number of health-related problems by means of toxic effects they may spark in the human body.
2. Microplastics tend to absorb contaminants, which may include persistent organic pollutants (POPs). The additional contaminants entering into the body along with microplastics create yet another pathway to physical and toxic influences on human health.

A very wide variety of additives as well as of performance/quality-enhancing chemicals, such as plasticizers, light and heat stabilizers, antioxidants, flame retardants, etc., are added during the process of polymer production (Kannan and Vimalkumar 2021). There are as many as 85,000 chemicals which have been produced, out of which thousands are endocrine-disrupting chemicals (EDCs—the chemicals that can mimic the shape of the hormone and fit into its receptor) and phthalates (e.g., vinyl flooring, lubricating oils, and personal-care products like soaps, shampoos, hair sprays, etc.) in the plastics which eventually dump into oceans. Quoting endocrinologist Dr. Ivone Mirpuri, Barrett (2019) writes that these chemicals also trigger serious ailments like obesity, diabetes, heart attacks, infertility, sex malformation, cancer, behavioral changes, and neurological disorders including Attention Deficit Hyperactivity (ADH).

Leached from the plastisphere, these additives mix with marine water where they become part and parcel of aquatic life and keep on magnifying their concentration along trophic hierarchy. They find passage to marine fauna through microplastics. Approximately 35 to 917 tons of additives are released into marine waters annually (Suhrhoff et al. 2016), which include the commonly known and widely used plastic additives like bisphenol A (BPA), nonylphenol, phthalate esters (PAEs), and polybrominated diphenyl ethers (PBDEs).

Most of the additives are present in plastics/microplastics at the rate of 20% (w/w). A few plasticizers, however, might be up to quite higher amounts, such as 70% (w/w) (Hermabessiere et al. 2017; Kannan and Vimalkumar 2021). Wiesinger et al. (2016) reported that as many as 10,000 chemicals were used as plastic monomers, additives, and processing aids. Out of these, more than 2400 have been detected to be of concern. More than 30,000 substances have been tested for their plasticizing properties. At present, nearly 50 of these substances are being used for commercial purpose. Approximately 85% of the plasticizers are employed in flexible polyvinyl chloride (PVC) applications in Europe. As documented by Kannan and Vimalkumar (2021), plasticizers in huge amounts, about 7.5 million tons, are used worldwide annually, out of which the European countries consume about 1.35 million tons. DEHP (diethylhexyl phthalates) account for 37% of the total plasticizer consumption.

A number of heavy metals, as per their broad definition given by Appenroth (2010), are used as additives to polymer products to improve the quality and attractiveness of the plastic products during their manufacturing process. The added metals serve as inorganic pigments or colorants (Al, Ba, Cd, Co, Cr, Mn, Pb, Ti, Zn), flame-retardants (Al, Sb, Br, Zn), heat stabilizers (Al, Cd, Pb, Zn), UV stabilizers (Al, Ba, Cd, Pb, Sn, Ti), anti-slip agents (Zn), biocides (As, Cu, Hg, Sb, Sn), etc. These metal-based additives have been known as the potential causal agents of serious human health problems including cytotoxicity, genotoxicity, breast cancer, DNA methylation, neurodegenerative disorder, brain damage, etc. and their toxicity may even lead to death (Table 4.3).

TABLE 4.3
Heavy metals used in additives in plastic products and their impact on human health.

Heavy metals	Additives	Polymer types	Effects on human health
Aluminum (Al)	Stabilizers, flame retardants, inorganic pigments	PE, PBT, PET, PVC	Metal estrogen, breast cancer
Antimony (Sb)	Biocides, flame retardants	Many types of polymers	Metal estrogen, breast cancer
Arsenic (As)	Biocides	LDPE, polyesters, PVC	Gastrointestinal damage; congenital diabetes; and skin, lung, liver, kidney, bladder cancer; death
Barium (Ba)	UV stabilizers, inorganic pigments	PVC	Metal estrogen; metabolic, neurological, and mental disorders; cardiovascular and kidney disorders; breast cancer
Bromine (Br)	Flame retardants	PBT, PE, PP, PS	Genotoxicity, apoptosis
Cadmium (Cd)	Inorganic pigments, heat stabilizers, UV stabilizers	PVC	Alteration in calcium, phosphorus, and bone metabolism; osteomalacia and bone fracture incidents in postmenopausal women; lipid peroxidation; promotion of carcinogenesis; DNA methylation, apoptosis

(Continued)

TABLE 4.3
(Continued)

Heavy metals	Additives	Polymer types	Effects on human health
Chromium (Cr)	Inorganic pigments	PE, PP, PVC	Allergic reactions; nasal septum ulcer; severe gastrointestinal, hepatic, respiratory, hematological, cardiovascular, renal, and neurological effects; possibly death
Cobalt (Co)	Inorganic pigments	PET bottles	ROS formation, cardiovascular and endocrine deficits, hearing and visual impairments
Copper (Cu)	Biocides	–	ROS formation, inducing DNA strand breaks and oxidation
Lead (Pb)	Inorganic pigments, heat stabilizers, UV stabilizers	PVC, All plastic types imparted red color	Hemoglobin deficiency leading to anemia, hypertension, miscarriages, infertility, oxidative stress, disruption of nervous system, cell damage, brain damage
Manganese (Mn)	Inorganic pigments	–	Neurodegenerative disorder
Mercury (Hg)	Biocides	PU	Mutagen/carcinogen, disruption of DNA molecular structure, brain damage
Tin (Sn)	UV stabilizers, biocides	PU foam, PVC	Nausea, skin rashes, vomiting, diarrhea, abdominal pain, headaches, palpitations, potential clastogens, metal estrogen, breast cancer
Titanium (Ti)	Inorganic pigments, UV stabilizers	PVC	Cytotoxicity on human epithelial lung and colon cells
Zinc (Zn)	Inorganic pigments, heat stabilizers, flame retardants, anti-slip agents	PE, PP, PVC	–

Source: Campanale et al. (2020).

Note: LDPE—low-density polyethylene; PBT— polybutylene terephthalate; PE—polyethylene; PET—polyethylene terephthalate; PP—polypropylene; PS—polystyrene; PU—polyurethane; PVC—polyvinyl chloride; ROS—reactive oxygen species.

The additives are also released from the plastic material during recycling and recovery processes as well as from various products manufactured from recyclates (Hahladakis et al. 2018), and thus, lead to human exposure to various health problems through various pathways.

4.6 MICROPLASTICS AS A POTENTIAL SOURCE OF HUMAN PATHOGENS IN THE MARINE ENVIRONMENT

The plastisphere is the source of a diversity of organisms dominated by microorganisms, many of which are widely recognized as vectors for the propagation of metals, antibiotics, and pathogens. Due to their very slow decomposition rates, microplastics keep serving as long-enduring "active surfaces"—attracting a variety of microbial communities for biofilm formation and gene exchange, and absorbing heavy metals, pesticides, and several other toxic substances.

Microplastics and nanoplastics are transported along with organic contaminants via sorption or desorption, a process known as the "Trojan-Horse Effect" (Zhang and Xu 2020). It is on account of this Trojan-Horse Effect that potential health risks of the microplastics may be significantly enhanced. Attachment of hazardous pathogens to plastic wastes was first of all reported by Masó et al. (2003), suggesting "drifting plastic debris as a potential vector for microalgae dispersal". From the marine environment to dispersal into the aerial environment, microplastics do not just follow a trophic food chain. They can also spread into terrestrial and aerial environments through ontogenic transference pathways. Conducting experiments on *Culex* mosquito life stages, Al-Jaibachi et al. (2018) proved that microplastics can make ontogenic transference from the larva to the pupa stage, and then onto the next adult form of the terrestrial stage. This transference depends on particle size: the smaller the microplastic size the faster the transference. The conclusion is that any organism feeding on the terrestrial life phase of aquatic insects can be influenced by the microplastics present in the water.

Microplastics giving refuge to a variety of microorganisms can spread antibiotic resistance, affecting aquatic ecology and evolution of aquatic life. The human species cannot be excluded from the microplastic impacts. Arias-Andres et al. (2018) in their study investigated that horizontal gene movement between phylogenetically distinctive microorganisms prospering on microplastic surfaces occurs more rapidly than between free-living microorganisms.

Among the microbes inhabiting microplastic surfaces are pathogenic ones for humans. *Vibrio* spp., a bacteria often referred to as containing pathogenic strains adversely affecting human health, occupies microplastic surfaces. In the marine environment in the North Sea and Baltic Sea, *Vibrio parahaemolyticus*, a potentially pathogenic strain, has been discovered on PE, PS, and PP microplastics (Kirstein et al. 2016). *Vibrio parahaemolyticus* causes various types of illness, such as gastroenteritis, wound infections, and sepsis. Bacterial gastroenteritis in Asia, especially in Japan, is commonly caused by this strain of bacteria. In USA, outbreaks of *V. parahaemolyticus* are on the rise (Rezny and Evans 2021). *Vibrio cholerae*, another bacterial strain found on microplastics, is the etiological agent of cholera, one of the frequently discussed epidemic diseases because of its spread at a faster rate in human settlements supplied with contaminated water.

Microplastics serve as vectors for metals, antibiotics, hazardous chemicals, pathogens, dinoflagellates, etc. and in ballast waters these play a detrimental role in spreading multiple drug-resistant human pathogens (Naik et al. 2019). Dinoflagellates, the monophyletic group of single-celled eukaryotes and marine plankton, form harmful algal blooms (HABs). The HABs and bacterial disease outbreaks can pose serious risks to public health worldwide.

Microplastic particles in the ocean ecosystem are the principal sources of anthropogenic contamination. Human health risks emanating from microplastics are pervasive, recalcitrant, and may continue unceasingly in the times to come (Bowley et al. 2021). Microplastics have the potential to spread human and animal pathogens into new and distant areas and create new targets for various diseases. Pathogenicity and the virulence potential of the pathogens thriving on microplastics, however, are yet to be understood in their entirety. Scientific consensus is growing that microplastic particles can serve as vectors for the proliferation of antimicrobial-resistance genes (Bowley et al. 2021). Ecological principles involving interconnectedness among environmental factors, organisms, and living processes help us understand the consequences of everything else occurring in nature.

Biofilms are especially "beneficiary" for the pathogens to prosper actively and profusely. Unique ecocoronas, in addition to biofilms, rapidly develop on microplastic surfaces and play a crucial role in support of the unique plastisphere communities (Moons et al. 2009; Bowley et al. 2021). The polymer type and environmental factors prevailing in marine ecosystems determine the process of ecocoronas rising along with biofilm formation. Living within a biofilm can make microbes emerge more virulent than they otherwise ought to be, as Lyons et al. (2010) investigated through their experiments with culturable infectious *Vibrio* spp. According to the researchers, considerably faster rates of metabolic response and functional diversity are exhibited by the aggregate-associated microbial communities rather than by the aggregate free-living ones, and these organic aggregates were thought to represent as "microscopic islands". Another clue is the greater frequency of horizontal gene transfer among microbes within microplastic biofilms than among free-living microorganisms (Arias-Andres et al. 2018). Such an enhanced state of microbial living, as confirmed through a probe on *Vibrio vulnificus*, *V. diabolicus*, and *V. parahaemolyticus* (Klein et al. 2018), may lead to the formation of what is referred to as "pathogenicity islands".

Antimicrobial resistance bacteria (ARB) prevailing in concentrations 100 to 5000 times higher on a microplastic surface than in the adjacent marine environment have been brought to light by Zhang et al. (2020b) who suggest that the microplastics are toxic contaminants giving way to the enrichment of superbugs. Some of the interesting inferences stemming from the comprehensive study of Yang et al. (2019) exploring plastics in the marine ecosystem as a pool for antibiotic and metal resistance genes are:

1. Diversity of resistance genes is not influenced by plastic size.
2. Plastics in marine ecosystems serve as reservoir for antibiotic resistance genes (ARGs) and metal resistance genes (MRGs).
3. Composition of the microorganisms' community is a determinant of the ARG profile but not of MRG.
4. MRGs have higher abundance than ARGs.
5. Bacteria of the family Flavobacteriaceae are potential hosts for ARGs and MRGs.

Microplastics in the marine ecosystem, in essence, serve not only as carriers for the pathogens, but also contribute to enrich pathogenic strains within pathogenicity islands, and possess other critical antimicrobial properties following the pathway of horizontal gene transfer.

Various pathogens may reach humans through seafood/aquatic food. For example, *Vibrio* spp. are responsible for an ailment in bivalves, frequently leading to en masse mortality of larvae and within adult populations (Beaz-Hidalgo et al. 2010; Solomieu et al. 2015). PS microplastics demonstrate different bacterial communities than PE and PP microplastics. Oyster pathogen *Vibrio splendidus* detected on microplastics (Frère et al. 2018) can transfer to humans incorporating seafood in their diets. Microplastics transferring harmful pathogens through edible bivalves have also been detected (Bowley et al. 2021).

4.7 EFFECTS OF MICROPLASTICS ON HUMAN HEALTH

Human exposure to microplastics pollution leads to serious health problems: metabolic disturbances, oxidative stress, inflammatory lesions, increased uptake or translocation, neurotoxicity, and enhanced cancer risks (Rahman et al. 2021).

Microplastic particles through ingestion by means of food and water enter into the intestines where they can disturb the intestinal microenvironment affecting an individual's health. The particles can cause oxidative damage as well as inflammation in the gut, as many studies reviewed by Huang et al. (2021) would reveal. Their review also highlights the negative consequences of the microplastics' presence in the gut, such as microbial disorders, gut epithelium damage, decrease of the mucus layer, immune cell toxicity, etc.

Size, shape, and abundance of plastic pieces influence uptake in lower trophic fauna, microfibers being the most harmful (Wright et al. 2013). This nature of microplastics also applies in humans. A study comparing cytotoxicity and efflux pump inhibition capability of different sizes of PS microplastic sizes in the human adenocarcinoma Caco-2 cells conducted by Wu et al. (2019) brings out many interesting facts. Caco-2 cells, an immortalized cell line of human colorectal adenocarcinoma cells primarily used as a model of the intestinal epithelial barrier (Lea 2015), experienced low toxicity effects exerted by different PS microplastic sizes on cell viability, oxidative stress, and membrane integrity. The mitochondrial membrane potential, however, underwent disruption with the impact of the 5 μm size, the impact being greater than that of 0.1 μm. The overall effects of the two different sizes of PS microplastics are shown in Figure 4.4. The information on the toxicity of varying microplastic sizes in human intestine cells provided by this experiment could be useful in the assessment of the risks associated with PS microplastics.

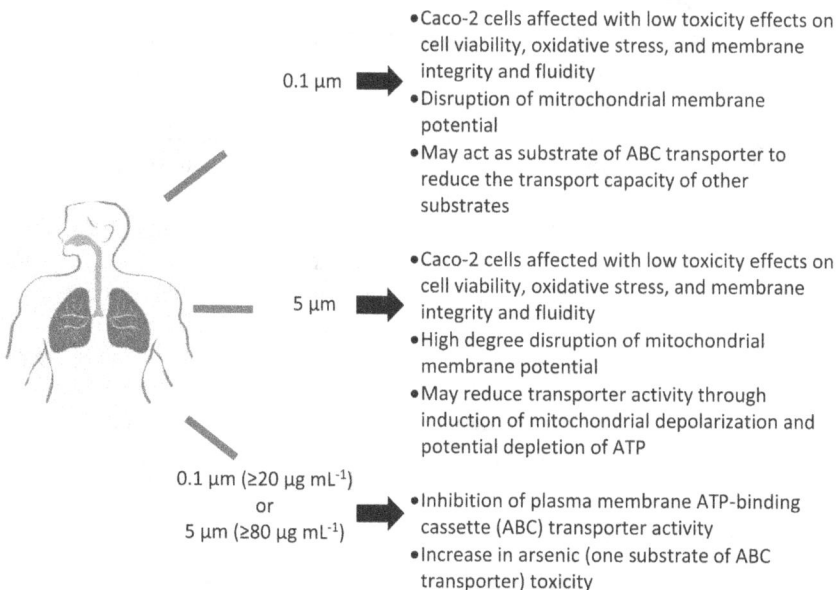

FIGURE 4.4 Effects of two different polystyrene microplastic sizes on human Caco-2 cells.

Source: Based on Wu et al. (2019).

Forte et al. (2016) in their experiment on PS nanoparticles internalization in human gastric adenocarcinoma cells demonstrated that in comparison to NP100, NP44 accumulate more rapidly and with greater efficiency in the cytoplasm of gastric adenocarcinoma. Both of the PS nanoparticles (NP44 and NP100) expressed an internalization mechanism that depends on energy together with clathrin-mediated endocytosis pathway. The PS nanoparticles were also found to influence cell viability, cell morphology, and inflammatory gene expression.

All biological systems, including those of humans, suffer from environmental exposure to microplastics in a number of ways with potentiality to cause particle toxicity (Prata et al. 2020):

1. Inflammatory lesions, oxidative stress, and enhanced uptake or translocation.
2. Disruption of immune function and neurotoxicity, and other grave consequences.
3. Exacerbated and worsened health problems because of the immune system's inability to get rid of the synthetic particles.
4. Chronic inflammation and increased risk of neoplasia.

Critical factors that influence the extent of an individual's health problems are concentrations of the nanoplastics to which humans are exposed, properties of the particles, contaminants absorbed along with the particles, tissues the particles make entry into, and susceptibility of the individuals exposed (Prata et al. 2020).

Workers exposed to microplastic-laden dust and air suffer from dyspnea and interstitial inflammatory responses. Susceptible persons might be at risk of developing similar lesions even when microplastic concentrations in the breathing air are low (Prata 2018). When inflammation becomes chronic, it might pave the way to very serious health problems.

Significant human health problems occurring due to microplastics and aided by nanoplastics entering into human body through various sources via inhalation, ingestion, and dermal contact are summarized in Figure 4.5.

Huang et al. (2022) investigated the transgenerational effects of nanoplastics in mice. Exposed to varying 100 nm PS nanoplastics concentrations (0.1, 1, and 10 mg L^{-1}) during gestation and lactation of 100 nm-sized PS particles, the mice under experiment were observed with impaired offspring development. No effects on the small intestine and kidneys of the pups of exposed mice were observed, but considerable impact of PS nanoplastics concentration on liver morphology, reduced liver size and weight, induction of oxidative stress, and promotion of an inflammatory response and glycometabolism disorder were investigated. Symptoms of male reproductive toxicity, including decreased sperm count and testicular oxidative injury, were also observed. Such transgenerational effects, likely to occur in humans, are yet to be established through research.

4.8 EPIDEMIOLOGY

Microplastics are used as carriers of medications into human body tissues. It is often hypothesized that the additives and contaminants absorbed by microplastics would act in a similar way in human tissues as these do in the marine environment (Smith

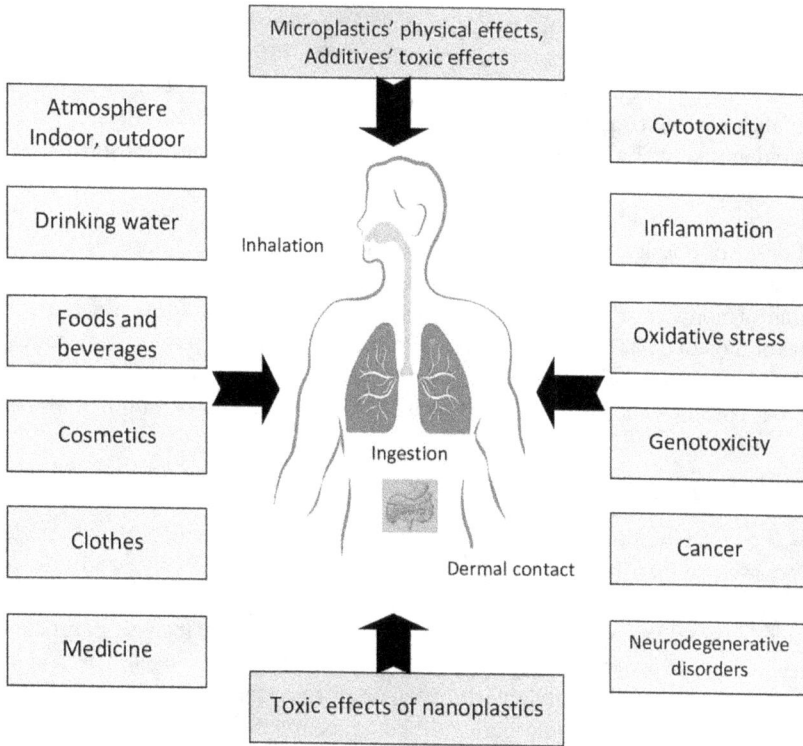

FIGURE 4.5 Significant human health problems generated by microplastics entering into the human body through various sources.

et al. 2018). Biomonitoring by measuring the concentration of certain harmful chemicals in human tissues provides evidence that they have entered through microplastic contamination. These are the same chemicals which are used in the manufacturing of plastic items (Talsness et al. 2009; Thompson et al 2009). The biomonitoring approach has found phthalates, bisphenol A (BPA), and many other additives in human populations and geographical area and age as the major factors responsible for the differences in data generated during studies (Thompson et al. 2009). Some phthalates, like diethylhexyl phthalates (DEHP), in addition to dust and foods, also reach human tissues through drugs.

How microplastics interact in human body tissues is yet to be understood fully and precisely. It is less likely that no adverse interaction would occur if some foreign elements make their entry into biological systems of the body. However, without adequate and extensive epidemiological studies, significant impacts of microplastic interactions in body tissues at the population level are difficult to be detected (Smith et al. 2018). A significant correlation between BPA level in urine, and cardiovascular disease and type 2 diabetes has been observed earlier by Melzer et al. (2010) in the adult US population. BPA ($C_{15}H_{16}O_2$ or $(CH_3)_2C(C_6H_4OH)_2$)), first synthesized in the 1930s as a synthetic estrogen, is used as a monomer to manufacture polycarbonate plastic and as an epoxy resin lining foods and beverage cans (Galloway 2015). These plastics are used to make beverage containers and several other items. Effects of BPA on human health in lower environmental concentrations are not known, but in experimental laboratory animals it has been observed affecting reproductive systems (CDC 2017).

Epidemiological and laboratory studies elicit evidence that exposure to BPA at levels found in the urine samples of the general population (about 0.2–20 ng ml^{-1}) is associated with adverse human health effects that include cardiovascular disease and onset of obesity, as documented by Galloway (2015). Low-dose BPA exposure in humans occurs due to microplastic particles and low- and high-dose exposure by means of non-microplastics sources: via inhalation of BPA contaminated air and consumption of contaminated foods (Smith et al. 2018). Microplastics may be responsible for exerting localized particle toxicity. However what is of greater concern is whether chronic exposure producing a cumulative effect occurs (Smith et al. 2018).

4.9 CONTROL MEASURES

Looking at the multiple risks and treacherous implications of microplastics pollution for ecosystems and human health, we need to consciously evolve and effectively implement control measures to mitigate this mushrooming predicament. Of immediate need is chalking out a national level policy to curb increasing rise in production, distribution, and overwhelming use of single-use plastic products. The single-use plastic products readily available in the market have rapidly become an unabated nuisance and they are contaminating all components of the environment more than other plastic types.

Extensive plastic usage in agriculture and plastic dumping in agro-ecosystems is posing a dual threat: food insecurity, and poor public health. Ecological alternatives to throwaway items and transforming hazardous wastes into useful ones for ameliorating agro-ecosystem health must be central to the strategies for a healthy, vibrant, and resilient food production system and food security.

In order to counter adverse effects of plastics and microplastics on nature, livelihoods, and public health, many countries have adopted legal measures. According to a United Nations Environment Program (UNEP) report on *Legal Limits on Single-Use Plastics and Microplastics*, 127 of the world's countries have instituted some kind of regulation on single-use plastics. These include bans on single-use plastics (e.g., plastic products such as bags, cups, plates, straws, packaging materials, etc. by 27 countries); taxes levied on the production of plastic items by 27 other countries; charging of consumer fees for plastic bags by 30 countries; inclusion of the features of extended producers' responsibility for plastics items within legislation by 43 countries; and mandate for extended producers' responsibility for single-use plastic products. Apart from these legislative measures relating to single-use plastics, eight countries, namely, Canada, USA, France, Italy, Republic of Korea, United Kingdom (Great Britain and Northern Ireland), Sweden, and New Zealand, have banned microbeads through law or regulations (UNEP 2018).

However, despite such legislations, companies continue producing an estimated five trillion single-use bags every year (Kurtz and Sample 2021). Such regulations do not seem to be enough to contain microplastics pollution and its known and unknown repercussions on human health. The laws needed to effectively curb and finally mitigate this significant global crisis need to cover the entire globe by instituting strict specific regulations at local (provinces/states, municipalities), regional (continental, international groups, forums, etc.), and global levels. Since numerous evidence is emerging declaring microplastics as one of the most pressing environmental and socioeconomic issues with the potential to attain pandemic proportions, the crisis needs to be dealt with like UN agendas on deserts, climate change, and health.

Health risks are ever higher for the communities living close to plastic production facilities and other petrochemical plants. These risks include high rates of cancer and other serious health hazards (Kurtz and Sample 2021). Therefore, such production facilities and petrochemical plants must be as far away from a community as possible which is necessary for people to avoid the direct impacts of the microplastics pollution.

Let us apply what can be called the 4 Rs Principle—Reduce, Reuse, Replace, and Recycle—to disallow plastics, microplastics, and nanoplastics from becoming a global menace:

- **Reduce** dependence on plastic material as much as you can, and reduce the plastic materials to the minimum; use them only when they are absolutely necessary. Unnecessary plastic items, such as single-use poly bags, may completely be avoided. Production of such items must also be declared "obscene".
- **Reuse** high-quality plastics for non-food/non-consumable items so that the frequency of throwing away plastic items is reduced to a minimum.
- **Replace** the plastic items by those produced using safe raw material (e.g., wood, stones, cloths, jute, metals, etc.) which are enduring and safe from a health point of view. Provision of incentives to use biodegradable and/or replaceable alternatives would be quite promising. Food packaging materials should be replaced with biodegradable materials.
- **Recycle** the plastics to the maximum possible extent. For this, incentives should be given to those who undertake this task following environmental concerns. In the process of recycling, emission of toxic substances and contamination of environmental components and recycled products must be avoided.

A worldwide policy encompassing country-wise specific legislation to address all issues pertaining to plastics and microplastics is an imperative of our times. This is also essential to meet global targets under Sustainable Development Goal 14 that seeks to ensure the conservation and sustainable use of oceans.

4.10 SUMMARY

Exposure of most of the planet's life, not just human's, to microplastics pollution becomes almost inevitable looking at the recorded microplastic presence in all the planet's ecosystems. Air—indoor as well as outdoor—has microplastics which are accidentally inhaled by humans. Microplastics are consumed along with water, bottled as well as tap water, and ingested with foods. Foods could be contaminated by microplastics *in situ*, like seafood, or during food processing and packaging. A variety of market-based drinks and beverages, such as wine, beer, bottled tea, energy drinks, etc. are also a rich source of microplastics. Baby feeding bottles are another source introducing microplastics into infants' bodies. Textile fibers, masks containing plastics, some medicines, medical equipment, plastic-made domestic gadgets, etc. constitute the other category of the plastics contributing to human exposure to microplastics pollution. There is a very long list of cosmetics widely used across the world that contain microbeads, the microplastic particles generally less than 1 mm in size used as abrasives in cosmetics. A few examples are hair bleaches, hair colorants, deodorants, nail polishes, lipsticks, sunscreens, shower gels, scrubs, soaps, shampoos, skin creams, etc. Very small microplastic particles can also make entry into the human body through dermal contact.

Seafoods are likely to be contaminated *in situ* as the oceans and seas are the ultimate dumping grounds of plastics. Table salt from Asian countries recorded higher microplastic concentrations from the countries outside the Asian continent. Seaweed, fish, bivalves, crustaceans, chickens, salt, sugar, and honey all have been found containing microplastics. Foods packed in plastic wraps or plastic containers are invariably contaminated by microplastics. Commonly found microplastics in the packaged food items emanate from PE, PET, PUR, and PS, PVC, and PP, polyamide, PMMA, and styrene acrylate.

Microplastics may contain nearly 4% of additives which can also adsorb toxic pollutants in their surrounding environment. The contaminants as well as the additives may be organic and inorganic types. Fish have been found containing high concentrations of microplastics in their stomach and intestines. Higher exposure risks are there for those who consume crustaceans and bivalve mollusks as the digestive tracts of this seafood are consumed. Microplastics may attract accumulation of toxic chemicals, like PAHs and PCBs, in high concentrations. BPA, a compound used in food packaging, may leave its residues on the foods. Consumers incorporating the foods with microplastics contaminated with chemicals would have the chemical pollutants transferred to their tissues.

More toxic than microplastics, nanoplastics have been detected in both polar regions of the planet and they are now pervasive around the world. Information on toxicokinetics gives an idea only about their absorption and distribution, but there is no information about their metabolism and excretion. Nanoplastics can make entry into body cells and may cause ill effects on human health. They may be responsible for inflammation, cellular barriers, or cross the blood-brain barrier and the placenta. The several disorders caused by nanoplastics are cardiopulmonary responses, changes in endogenous metabolites, genotoxicity, inflammation, oxidative stress, effects on the absorption of nutrients, changes in gut microflora, and reproduction disorders, etc.

The "safety" in the food security definition is compromised when the food is contaminated by undesirable plastic particles. The mess of the plastics and microplastics in the marine environment leads to decline in secondary productivity, a state that impacts humans' seafood-dependent food security. Adverse impacts of plastic pollution on livestock and livestock-dependent communities, such as in poor countries, lead to an increase in food insecurity and health problems.

Single-use plastics seem to have become like an invasive species in water bodies. Most microplastics go in and out of most organisms. Laboratory experiments reveal that microplastics in high amounts and under specific conditions can induce physical and chemical toxicity. Negative effects of microplastics on food intake, growth, reproduction, and survivability, etc. have been observed. Serious health risks to humans in the communities living close to plastic manufacturing facilities or petrochemical plants are also evident and such risks disproportionately impact vulnerable populations. Colonization of microplastics marks their effects at the population level; size, shape, and abundance have influence on microplastic uptake. Ingestion of microplastics leads to microplastic toxicity which happens through three possible pathways, namely ingestion stress, additive leakage, and absorption of contaminants by microplastics. Microplastics ingestion causes stress in response to energy expenditure required for egestion, physical blockage, and a sort of false satiation. As the microplastics are undesirable foreign elements making entry into the body, the digestive and excretory systems need to get rid of them, a process that requires expenditure of extra energy, thus putting an individual under constant stress and possible physiological disturbances.

A very wide variety of additives as well as performance/quality-enhancing chemicals, such as plasticizers, light and heat stabilizers, antioxidants, flame retardants, etc., are added during the process of polymer production. These chemicals also trigger serious ailments like stunted fertility and sex malformations, obesity, diabetes, heart attacks, cancer, and behavioral and other neurological problems. A number of heavy metals like Al, Ba, Cd, Co, Cr, Mn, Pb, Sb, Sn, Ti, Zn, etc. are used as additives to polymer products to improve the quality and attractiveness of the plastic products during their manufacturing process. These metal-based additives have been known as the potential causal agents of serious human health problems including cytotoxicity, genotoxicity, breast cancer, DNA methylation, neurodegenerative disorders, brain damage, etc. and their toxicity may even lead to death.

The Plastisphere is the source of a diversity of organisms dominated by microorganisms many of which are widely recognized as vectors for the expansion of metals, antibiotics, and human pathogens. Microplastics harboring a variety of microorganisms can spread antibiotic resistance affecting aquatic ecology and evolution of aquatic life. Horizontal gene transfer between phylogenetically distinctive microbes perpetuating on microplastic surfaces is far faster than between free-living microbes. Among the microbes inhabiting microplastic surfaces are pathogenic ones for humans. *Vibrio* spp., a bacteria often referred to as containing pathogenic strains adversely affecting human health, occupies microplastic surfaces. *Vibrio parahaemolyticus* causes various types of illness, such as gastroenteritis, wound infections, and sepsis. *Vibrio cholerae* is the etiological agent of cholera, an epidemic disease of significant public health importance because of its very rapid spread in areas with supplies of contaminated water. Microplastics serve as a source of vector for metals, antibiotics, toxic chemicals, pathogenic bacteria, and dinoflagellates. Dinoflagellates form HABs. The HABs and bacterial disease outbreaks may pose a serious threat to worldwide public health. Significantly higher rates of metabolic response and functional diversity are exhibited by the aggregate-associated microbial communities rather than by the aggregate free-living ones and these organic aggregates were thought to represent "microscopic islands".

Human exposure to microplastics can cause serious health problems, such as toxicity, oxidative stress, inflammatory lesions, and enhanced uptake or translocation, metabolic disturbances, neurotoxicity, and enhanced cancer risks.

Biomonitoring by measuring the concentration of certain harmful chemicals in human tissues provides evidence that they have entered through microplastic contamination. These are the same chemicals which are used in the manufacturing of plastic products. The biomonitoring approach has found phthalates, BPA, and many other additives in human populations.

Epidemiological and laboratory studies elicit evidence that exposure to BPA at levels found in the urine samples of the general population (about $0.2–20$ ng ml^{-1}) is associated with adverse human health effects that include cardiovascular disease and onset of obesity.

Looking at the multiple risks and implications of microplastics pollution for ecosystems and human health, we need to consciously evolve and effectively implement control measures to mitigate this mushrooming dilemma. Of immediate need is chalking out a national level policy to curb increasing rise in production, distribution, and overwhelming use of single-use plastic products. These use-and-throw-away type plastic products have emerged as an unabated nuisance and they contaminate all components of the environment more than the other plastic types do. Let us apply what can be called the 4 Rs Principles—Reduce, Reuse, Replace, and Recycle—to disallow plastics, microplastics, and nanoplastics from becoming a global menace. A global policy encompassing country-wise specific legislation to address all issues pertaining to plastics and microplastics is an imperative of our times. Protection of the oceans from burgeoning microplastic pollution is also essential to meet global targets under Sustainable Development Goal 14 that seeks to ensure the conservation and sustainable use of oceans.

REFERENCES

Al Naggar, Y., Brinkmann, M., Sayes, C. M., Al-Kahtani, S. N., Dar, S. A., El-Seedi, H. R., Grünewald, B. and Giesy, J. P. 2021. Are honeybees at risk from microplastics? *Toxics*, 9(5):109. doi: 10.3390/toxics9050109.

Al-Jaibachi, R., Cuthbert, R. N. and Callaghan, A. 2018. Up and away: Ontogenic transference as a pathway for aerial dispersal of microplastics. *Biology Letters*, 14:20180479. http://dx.doi.org/10.1098/rsbl.2018.0479.

Appenroth, K. J. 2010. Definition of "heavy metals" and their role in biological systems. *Soil Biology*, 19:19–29. doi: 10.1007/978-3-642-02436-8_2.

Arias-Andres, M., Klümper, U., Rojas-Jimenez, K. and Grossart, H. P. 2018. Microplastic pollution increases gene exchange in aquatic ecosystems. *Environmental Pollution*, 237:253–261. doi: 10.1016/j.envpol.2018.02.058.

Barrett, T. 2019. Microplastic pollution "number one" threat to humankind. *Environment Journal*. https://environmentjournal.online/articles/microplastic-pollution-number-one-threat-to-humankind/

Beaz-Hidalgo, R., Balboa, S., Romalde, J. L. and Figueras, M. J. 2010. Diversity and pathogenecity of *Vibrio* species in cultured bivalve molluscs. *Environmental Microbiology Reports*, 2:34–43. doi: 10.1111/j.1758-2229.2010.00135.x.

Bergmann, M., Gutoww, L. and Klages, M. (eds.). 2015. *Marine Anthropogenic Litter*. Cham: Springer Open. 447pp.

Bessa, F., Barria, P., Neto, J. M., Frias, J. P., Otero, V., Sobral, P. and Marques, J. C. 2018. Occurrence of microplastics in commercial fish from a natural estuarine environment. *Marine Pollution Bulletin*, 128:575–584. doi: 10.1016/j.marpolbul.2018.01.044.

Bowley, J., Baker-Austin, C., Porter, A., Hartnell, R. and Lewis, C. 2021. Ocean Hitchhikers—assessing pathogen risks from marine microplastic. *Trends in Microbiology*, 29(2):107–116. doi: 10.1016/j.tim.2020.06.011.

Campanale, C., Massarelli, C., Savino, I., Locaputo, V. and Uricchio, V. 2020. A detailed review study on potential effects of microplastics and additives of concern on human health. *International Journal of Environmental Research and Public Health*, 17(4):1212. doi: 10.3390/ijerph17041212.

Carrington, D. 2022. Nanoplastic pollution found at both of Earth's poles for first time. *The Guardian*, January 21, 2022. www.theguardian.com/environment/2022/jan/21/nanoplastic-pollution-found-at-both-of-earths-poles-for-first-time

CDC (Centers for Disease Control and Prevention). 2017. www.cdc.gov/biomonitoring/BisphenolA_FactSheet.html#.

Cheung, P. K. and Fok, L. 2016. Evidence of microbeads from personal care products contaminating the sea. *Marine Pollution Bulletin*, 109(1):582–585. doi: 10.1016/j.marpolbul.2016.05.046.

Cox, K. D., Covernton, G. A., Davies, H. L., Dover, J. F., Juanes, F. and Dudas, F. 2019. Human consumption of microplastics. *Environmental Science and Technology*, 53(12):7068–7074. doi: 10.1021/acs.est.9bo1517.

Critchell, K. and Hoogenboom, M. O. 2018. Effects of microplastic exposure on the body conditions and behaviour of planktivorous reef fish (*Acanthochromis polyacanthus*). *PLoS ONE*, 13(3): e0193308. doi: 10.1371/journal.pone.0193308.

Dick Vethaak, A. and Legler, J. 2021. Microplastics and human health. *Science*, 371(6530). doi: 10.1126/science.abe5041.

Dris, R. Gasperi, J., Rocher, V., Saad, M., Renault, N. and Tassin, B. 2015. Microplastic concentration in an urban area: A case study in Greater Paris. *Environmental Chemistry*, 12(5):592–599. doi: 10.1071/EN14167.

Dris, R., Gasperi, J., Mirande, C., Mandrin, C., Guerrouache, M. and Tassin, B. 2017. A first overview of textile fibers, including microplastics, indoor and outdoor environments. *Environmental Pollution*, 221:453–458. doi: 10.1016/j.envpol.2016.12.013.

Dris, R., Gasperi, J., Saad, M., Mirande, C. and Tassin, B. 2016. Synthetic fibers in atmospheric fallout: A source of microplastics in the environment? *Marine Pollution Bulletin*, 104(1–2):290–293. doi: 10.1016/j.marpolbul.2016.01.006.

Du, H., Huang, S. and Wang, J. 2022. Environmental risks of polymer materials from disposable face masks and linked to the Covid-19 pandemic. *Science of the Total Environment*, 815:152980. doi: 10.1016/j.scitotenv.2022.152980.

Elizalde-Velázquez, G. A. and Gómez-Oliván, L. M. 2021. Microplastics in aquatic environments: A review on occurrence, distribution, toxic effects, and implications for human health. *Science of the Total Environment*, 780:146551. doi: 10.1016/j.scitotenv.2021.146551.

European Food Safety Authority (EFSA). 2016. Presence of microplastics and nanoplastics in food, with particular focus on seafood. *EFSA Journal*, 14(6):4501. doi: 10.2903/j.efsa.2016.4501.

FAO. 1996. *World Food Summit: Rome Declaration on World Food Security and World Food Summit Plan of Action*, Rome. www.fao.org/3/w3613e/w3613e00.htm.

Forte, M., Iachetta, G., Tussellino, M., Carotenuto, R. Prisco, M. aDe Falco, M., Laforgia, V. and Valiante, S. 2016. Polystyrene nanoparticles internalization in human gastric adenocarcinoma cells. *Toxicology in Vitro*, 31:126–136. doi: 10.1016/j.tiv.2015.11.006.

Fournier, E., Etienne-Mesmin, L., Grootaert, C., Jelsbak, L., Syberg, K., Blanquet-Diot, S. and Mercier-Bonin, M. 2021. Microplastics in the human digestive environment: A focus on the potential and challenges facing in vitro gut model development. *Journal of Hazard Mater*, 415:125632. doi: 10.1016/j.jhazmat.2021.125632.

Frère, L., Maignien, L., Chalopin, M., Huvet, A., Rinnert, E., Morrison, H., Kerninon, S., Cassone, A.-L., Lambert, C., Reveillaud, J. and Paul-Pont, I. 2018. Microplastic bacterial communities in the Bay of Brest: Influence of polymer type and size. *Environmental Pollution*, 614–625. doi: 10.1016/j.envpol.2018.07.023.

Galloway, T. S. 2015. Micro- and nano-plastics and human health. In: Bergmann, M., Gutoww, L. and Klages, M. (eds.) *Marine Anthropogenic Litter*. Cham: Springer Open. 343–366. doi: 10.1007/978-3-319-16510-3_13.

Gouin, T., Roche, N., Lohmann, R. and Hodges, G. A. 2011. Thermodynamic approach for assessing the environmental exposure of chemicals absorbed to microplastic. *Environmental Science and Technology*, 45(4):1466–1472. doi: 10.1021/es1032025.

Hahladakis, J. H., Velis, C. A., Weber, R., Iacovidou, E. and Purnell, P. 2018. An overview of chemical additives present in plastics: Migration, release, fate and environmental impact during their use, disposal and recycling, *Journal of Hazard Mater*, 344:179–199. doi: 10.1016/j.jhazmat.2017.10.014.

Hermabessiere, L., Dehaur, A. Paul-Pont, I., Lacroix, C., Jezequel, R., Soudant, P. and Duflos, G. 2017. Occurrence and effects of plastic additives on marine environments and organisms: A review. Chemosphere, 182:781–793. doi: 10.1016/j.chemosphere.2017.05.096.

Hernandez, L. M., Xu, E. G., Larsson, H. C., Tahara, R., Maisuriya, V. B. and Tufenkii, N. 2019. Plastic teabags release billions of microparticles and nanoparticles into tea. *Environmental Science and Technology*, 53(21):12300–12310. doi: 10.1021/acs.est.9b02540.

Huang, T., Zhang, W., Lin, T., Liu, S., Sun, Z., Liu, F. Yuan, Y., Xiang, X., Kuang, H., Yang, B. and Zhang, D. 2022. Maternal exposure to polystyrene nanoplastics during gestation and lactation period and lactation induces hepatic and testicular toxicity in male mouse offspring. *Food and Chemical Toxicology*, 160:112803. doi: 10.1016/j.fct.2021.112803.

Huang, Z., Weng, Y., Shen, Q., Zhao, Y. and Jin, Y. 2021. Microplastic: A potential threat to human and animal health by interfering with the intestinal barrier function and changing the intestinal microenvironment. *Science of the Total Environment*, 785:147365. doi: 10.1016/j.scitotenv.2021.147365.

Kannan, K. and Vimalkumar, K. 2021. A review of human exposure to microplastics and insights into microplastics as obesogens. *Frontier of Endocrinology*, 12:724989. doi: 10.3389/fendo.2021.724989.

Karami, A., Golieskardi, A., Choo, C. K., Larat, V., Galloway, T. S. and Salamatinia, B. 2017. The presence of microplastics in commercial salts from different countries. *Science Reports*, 7(1):1–11. doi: 10.1038/srep46173.

Karbalaei, S., Hanachi, P., Walker, T. R. and Cole, M. 2018. Occurrence, sources, human health impacts and mitigation of microplastic pollution. *Environmental Science and Pollution Research*, 25(36):36046–36063. doi: 10.1007/s11356-018-3508-7.

Kedzierski, M., Lachat, B., Sire, O., Le, Maguer G., Le, Tilly V. and Bruzaud, S. 2020. Microplastic contamination of packaged meat: Occurrence and associated risks. *Food Package Shelf Life*, 24:100489. doi: 10.1016/j.fpsl.2020.100489.

Kim, J.-S., Lee, H.-J., Kim, S.-K. and Kim, H-J. 2018. Global pattern of microplastics (MPs) in commercial food-grade salts: Sea salt as an indicator of seawater MP pollution. *Environmental Science and Technology*, 52(21):12819–12828. doi: 10.1021/acs.est.8b04180.

Kirstein, I. V., Kirmizi, S., Wichels, A., Garin-Fernandez, A., Erler, R., Löder, M and Gerdts, G. 2016. Dangerous hitchhikers? Evidence for potentially pathogenic *Vibrio* spp. on microplastic particles. *Marine Environment Research*, 120:1–8. doi: 10.1016/j. marenvres.2016.07.004.

Klein, S., Pipes, S. and Lovell, C. R. 2018. Occurrence and significance of pathogenicity and fitness islands in environmental vibrios. *AMB Express*, 8:177. doi: 10.1186/ s13568-018-0704-2.

Koelmans, A. A., Nor, N. H. M., Hermsen, E., Kooi, M., Mintenig, S. M. and De France, J. 2019. Critical review and assessment of data quality: *Water Research*, 155:410–422. doi: 10.1016/j.waters.2019.02.054.

Kurtz, J. and Sample, D. 2021. To build food security, reduce plastic use. *IFPRI Blog*. www. ifpri.org/blog/build-food-security-reduce-plastic-use.

Kutralam-Muniasamy, G., Pérez-Guevara, F., Elizalde-Martínez, I. and Shruti, V. 2020. Branded milks–are they immune from microplastics contamination. *Science of the Total Environment*, 714:136823. doi: 10.1016/j.scitotenv.2020.136823.

Lea, T. 2015. Caco-2 cell line. *The Impact of Food Bioactives on Health*, 103–111. doi: 10.1007/978-3-319-16104-4_10.

Lebreton, L., Slat, B., Ferrari, F. Sainte-Rose, B., Aitken, J., Marthouse, R., Hajbane, S., Cunsolo, S., Schwarz, A., Levivier, A., Noble, K., Debeljak, P., Maral, H., Schoeneich-Argent, R., Brambini, R. and Reisser, J. 2018. Evidence that the great pacific garbage patch is rapidly accumulating plastic. *Nature Science Reports*, 8:46666. doi: 10.1038/ s41598-018-22939-w.

Li, D., Shi, Y., Yang, L., Xiao, L., Kehoe, D. K., Gun'ko, Y. K., Boland, J. J. and Wang, J. J. 2020. Microplastic release from the degradation of polypropylene feeding bottles during infant formula preparation. *Nature Food*, 1(11):746–754. doi: 10.1038/s43016-020-00171-y.

Li, W. C., Tse, H. and Fok, L. 2016. Plastic waste in the marine environment: A review of sources, occurrence and effects. *Science of the Total Environment*, 566:333–3349. doi: 10.1016/j.scitotenv.2016.05.084.

Liebezeit, G. and Liebezeit, E. 2013. Non-pollen particulates in honey and sugars. *Food Additives and Contaminants: Part A*, 30(12):2136–2140. doi: 10.1080/19440049.2013.843025.

Lwanga, E. H., Vega, J. M., Quej, V. K., de los Angeles Chi, J., Del Cid, L. S., Chi, C., Segura, G. E., Gertsen, S., Salánki, T., van der Ploeg, M., Koelmans, A. A. and Geissen, V. 2017. Field evidence for transfer of plastic debris along a terrestrial food chain. *Science Report*, 7(1):1–7. doi: 10.1038/s41598-017-14588-2.

Lyons, M. M., Ward, J. E., Gaff, H. Hicks, R. E., Drake, J. M. and Dobbs, F. C. 2010. Theory of island biogeography on a microscopic scale: Organic aggregates as islands for aquatic pathogens. *Aquatic Microbial Ecology*, 60:1–3. doi: 10.3354/ame01417.

Masó, M. Garcés, E., Pagès, F. and Camp, J. 2003. Drifting plastic debris as a potential vector for dispersing Harmful Algal Bloom (HAB) species. *Science March*, 67(1):107–111.

Materić, D., Ludewig, E., Brunner, D. Röckmann, T. and Holzinger, R. 2021. Nanoplastics transport to the remote, high-altitude Alps. *Environmental Pollution*, 288:117697. doi: 10.1016/j.envpol.2021.117697.

Mathalon, A. and Hill, P. 2014. Microplastic fibers in the intertidal ecosystem surrounding Halifax Harbor, Nova Scotia. *Marine Pollution Bulletin*, 81(1):69–79. doi: 10.1016/j. marpolbul.2014.02.018.

Melzer, D., Rice, N. E., Lewis, C., Henley, W. E. and Galloway, T. S. 2010. Association of urinary bisphenol A concentration with heart disease: Evidence from NHANES 2003/06. *PLoS ONE*, 5(1): e8673. doi: 10.1371/journal.pone.0008673.

Moons, P., Michiels, C. W. and Aertsen, A. 2009. Bacterial biofilms bacterial interactions in biofilms. *Critical Review Microbiology*, 35(3):157–168. doi: 10.1080/10408410902809431.

Naik, R. K., Naik, M. M., D'Costa, P. M. and Shaikh, F. 2019. Microplastics in ballast water as an emerging source and vector for harmful chemicals, antibiotics, metals, bacterial pathogens and HAB species: A potential risk to the marine environment and human health. *Marine Pollution Bulletin*, 149:110525. doi: 10.1016/j.marpolbul.2019.110525.

Naji, A., Nuri, M. and Vethaak, A. D. 2018. Microplastics contamination in mollusks from the northern part of the Persian Gulf. *Environmental Pollution*, 235:113–120. doi: 10.1016/j.envpol.2017.12.046.

Napper, I. E., Bakir, A., Rowland, S. J. and Thompson, R. C. 2015. Characterization, quantity and sorptive properties of microplastics extracted from cosmetics. *Marine Pollution Bulletin*, 99(1–2):178–185. doi: 10.1016/j.marpolbul.2015.07.029.

Neves, D., Sobral, P., Ferreira, J. L. and Pereira, T. 2015. Ingestion of microplastics by commercial fish off the Portuguese Coast. *Marine Pollution Bulletin*, 101(1):119–126. doi: 10.1016/j.marpolbul.2015.11.008.

Nor, N. H. M., Kooi, M., Diepens, N. J. and Koelmans, A. A. 2021. Lifetime accumulation of microplastic in children and adults. *Environmental Science and Technology*, 55(8): 5084–5096. doi: 10.1021/acs.est.0c07384.

Oßmann, B. E., Sarau, G., Holtmannspötter, H., Pischetsrieder, M., Christiensen, S. H. and Dicke, W. 2018. Small-sized microplastics and pigmented particles in bottled mineral water. *Water Research*, 141:307–316. doi: 10.1016/j.watres.2018.05.027.

Prata, J. C. 2018. Air-borne microplastics: Consequences to human health? *Environmental Pollution*, 234:115–126. doi: 10.1016/j.envpol.2017.11.043.

Prata, J. C., da Costa, J. P., Lopes, I., Duarte, A. C. and Rocha-Santos, T. 2020. Environmental exposure to microplastics: An overview on possible health effects. *Science of the Total Environment*, 702:134455. doi: 10.1016/j.scitotenv.2019.134455.

Rahman, A., Sarkar, A., Yadav, O. P., Achari, G. and Slobodnik, J. 2021. Potential human health risks due to nano- and microplastics and knowledge gaps: A scoping review. *Science of the Total Environment*, 757:143872. doi: 10.1016/j.scitotenv.2020.143872.

Reed, C. 2015. Dawn of the Plasticene age. *New Scientist*, 225:28–32. doi: 10.1016/S0262-4079(15)60215-9.

Rezny, B. R. and Evans, D. S. 2021. Vibrio parahaemolyticus. www.ncbi.nlm.nih.gov/books/NBK459164/

SAPEA (Science Advice for Policy by European Academies). 2019. *A Scientific Perspective on Microplastics in Nature and Society*. Berlin: SAPEA. 173pp. doi: 10.26356/microplastics.

Schymanski, D., Goldbach, C., Humpf, H.-U., and Furst, P. 2018. Analysis of microplastics in water by Micro-Raman Spectroscopy: Release of plastic particles from different packaging into mineral water. *Water Research*, 129:154–162. doi: 10.1016/j.watres.2017.11.011.

Seltenrich, N. 2015. New link in the food chain? Marine plastic pollution and seafood safety. *Environmental Health Perspectives*, 123(2): A41. doi: 10.1289/ehp.123-A34.

Senathirajah, K., Attwood, S., Bhagwat, G., Carbery, M., Wilson, S. and Palanisami, T. 2021. Estimation of the mass of microplastics ingested—A pivotal first step towards human health risk assessment. *Journal of Hazard Matter*, 404:124004. doi: 10.1016/j.jhazmat.2020.124004.

Sharma, S. and Chatterjee, S. 2017. Microplastic pollution, a threat to marine ecosystem and human health: A short review. *Environmental Science and Pollution Research*, 24(27):21530–21547. doi: 10.1007/s11356-017-9910-8.

Sharma, V. K. 2020. Polymers and microplastics: Implications on our environment and sustainability. In: Lu, Q. and Serajuddin, M. (eds.) *Emerging Technologies, Environment and Research for Sustainable Aquaculture*. London: Intech Open. doi: 10.5772/intechopen.89571.

Shruti, V., Perez-Guevara, F., Elizalde-Martinez, I. and Kutralam-Muniasamy, G. 2020. Toward a unified framework for investigating micro (nano) plastics in packaged beverages intended for human consumption. *Environmental Pollution*, 268:115811. doi: 10.1016/j.envpol.2020.115811.

Sillanpää, M. and Sainio, P. 2017. Release of polyester and cotton fibers from textiles in machine washings. *Environmental Science and Pollution Research*, 24(23): 19313–19321. doi: 10.1007/s11356-017-9621-1.

Simon, G-A. 2012. Food security: Definition, four dimensions, history. www.fao.org/fileadmin/templates/ERP/uni/F4D.pdf.

Singh, Vir. 2019. *Fertilizing the Universe: A New Chapter of Unfolding Evolution*. London: Cambridge Scholars Publishing. 285pp.

Singh, Vir. 2020. *Environmental Plant Physiology: Botanical Strategies for a Climate Smart Planet*. Boca Raton: Taylor and Francis (CRC Press). 216pp.

Smith, M., Love, D. C., Rochman, C. M. and Neff, R. A. 2018. Microplastics in seafood and the implications for human health. *Current Environmental Health Reports*, 5(3): 375–386. doi: 10.1007/s40572-018-0206-z.

So, W. K., Chan, K. and Not, C. 2018. Abundance of plastic microbeads in Hong Kong coastal water. *Marine Pollution Bulletin*, 133:500–505. doi: 10.1016/j.marpolbul.2018.05.066.

Solomieu, V. B., Renault, T. and Travers, M.-A. 2015. Mass mortality in bivalves and the intricate case of the Pacific oyster, *Crassostrea gigas*. *Journal of Invertebrate Pathology*, 131:2–10. doi: 10.1016/j.jip.2015.07.011.

Sorasan, C., Edo, C., González-Pleiter, M., Fernández-Piñas, F., Leganés, F., Rodríguez, A. and Rosal, R. 2021. Generation of nanoparticles during the photoageing of low-density polyethylene. *Environmental Pollution*, 289:117919. doi: 10.1016/j.envpol.2021.117919.

Stager, C. *Deep Future: The Next 100,000 Years of Life on Earth*. New York: Thomas Dunne Books.

Suhrhoff, T. J. and Scholz-Böttcher, B. M. 2016. Qualitative impact of salinity, UV radiation and turbulence on leaching of organic additives from four common plastics—a lab experiment. *Marine Pollution Bulletin*, 102(1):84–94. doi: 10.1016/j.marpolbul.2015.11.054.

Talsness, C. E., Andrade, A. J. M., Kuriyama, S. N., Taylor, J. A. and vom Saal, F. 2009. Components of plastic: Experimental studies in animals and relevance for human health. *Philosophical Transactions of the Royal Society B*, 364:2079–2096. doi: 10.1098/rstb.2008.0281.

Teng, J., Wang, Q., Ran, W., Wu, D., Liu, Y., Sun, S., Liu, H., Cao, R. and Zhao, J. 2019. Microplastic in cultured oysters from different coastal areas of China. *Science of the Total Environment*, 653:1282–1292. doi: 10.1016/j.scitotenv.2018.11.057.

Thompson, R. C., Moore, C. J., vom Saal, F. S., Swan, S. H. 2009. Plastics, the environment and human health: Current consensus and future trends. *Philosophical Transactions of the Royal Society B: Biological Sciences*, 364(1526):2153–2166. doi: 10.1098/rstb.2009.0053.

Toussaint, B., Raffael, B., Angers-Loustau, A., Gilliland, D., Kestens, V., Petrillo, M., Rio-Echevarria, I. M. and den Eede, G. 2019. Review of micro- and nanoplastic contamination in the food chain. *Food Additives and Contaminants: Part A*, 36(5):639–673. doi: 10.1080/19440049.2019.1583381.

Uddin, M. A., Afroj, S., Hasan, T., Carr, C., Novoselov, K. S. and Karim, N. 2021. Environmental impacts of personal protective clothing used to combat Covid-19. *Advanced Sustainable Systems*, 13:2100176. doi: 10.1002/adsu.202100176.

UNEP (United Nations Environment Program). 2018. *Legal Limits on Single-Use Plastics and Microplastics: A Global Review of National Laws and Regulations*. UNEP. 112pp.

Vianello, A., Jensen, R. L., Liu, L. and Vollertsen, J. 2019. Simulating human exposure to indoor airborne microplastics using breathing thermal manikin. *Science Reports*, 9(1): 1–11. doi: 10.1038/s41598-019-45054-w.

Wang, Y., Huang, J., Zhu, F. and Zhou, S. 2021. Airborne microplastics: A review on the occurrence, migration and risks to humans. *Bulletin of Environmental Contamination and Toxicology*, 1–8. doi: 10.1007/s00128-021-03180-0.

Wiesinger, H., Wang, Z. and Hellweg, S. 2016. Deep dive into monomers, additives and processing aids. *Environmental Science and Technology*, 55(13):9339–9351. doi: 10.1021/acs.est.1coo976.

Wright, S. L., Thompson, R. C. and Galloway, T. S. 2013. The physical impacts of microplastics on marine organisms: A review. *Environmental Pollution*, 178:483–492. doi: 10.1016/j.envpol.2013.02.031.

Wu, B., Wu, X., Liu, S., Wang, Z. and Chen, L. 2019. Size-dependent effects of polystyrene microplastics on cytotoxicity and efflux pump inhibition in human Caco-2 cells. *Chemosphere*, 221:333–341. doi: 10.1016/j.chemosphere.2019.01.056.

Xu, C., Zhang, B., Gu, C., Shen, C., Yin, S., Amir, M. 2020. Are we underestimating the sources of microplastic pollution in terrestrial environment? *Journal of Hazard Mater*, 400:123228. doi: 10.1016/j.jhazmat.2020.123228.

Yang, Y., Liu, G., Song, W., Ye, C., Lin, H., Li, Z. and Liu, W. 2019. Plastics in marine environment are reservoirs for antibiotic and metal resistance genes. *Environment International*, 123:79–86. doi: 10.1016/j.envint.2018.11.061.

Zarus, G. M., Muianga, C. and Pappas, R. S. 2020. A review of data for quantifying human exposures to micro and nanoplastics and potential health risks. *Science of the Total Environment*, 756:144010. doi: 10.1016/j.scitotenv.2020.144010.

Zhang, J., Wang, L. and Kannan, K. 2019. Polyethylene terephthalate and polycarbonate microplastics in pet food and feces from the United States. *Environmental Science and Technology*, 53(20):12035–12045. doi: 10.1021/acs.est.9b03912.

Zhang, K., Xiong, X., Hu, H., Wu, C., Bi, Y., Wu, Y., Zhou, B., Lam, P. K. S. and Liu, J. 2017. Occurrence and characteristics of microplastic pollution in Xiangxi Bay of three gorges reservoir, China. *Environmental Science and Technology*, 51(7):3794–3801. doi: 10.1021/acs.est.7b00369.

Zhang, M. and Xu, L. 2020. Transport of micro- and nanoplastics in the environment: Trojan-Horse Effect for organic contaminants. *Critical Reviews in Environment Science Technology*, 52(5):810–846. doi: 10.1080/10643389.2020.1845531.

Zhang, Q., Zhao, Y., Du, F., Cai, H., Wang, G. and Shi, H. 2020a. Microplastic fallout in different indoor environments. *Environmental Science and Technology*, 54(11):6530–6539. doi: 10.1021/acs.est.0c00087.

Zhang, Y., Lu, J., Wu, J., Wang, J. and Luo, Y. 2020b. Potential risks of microplastics combined with superbugs: Enrichment of antibiotic resistant bacteria on the surface of microplastics in mariculture system. *Ecotoxicology and Environmental Safety*, 187:109852. doi: 10.1016/j.ecoenv.2019.109852.

5 Biodegradation of Microplastics by Microbes

Microplastics pollution entering into Earth's marine ecosystems is slowly but steadily assuming formidable proportions. Since most of the planet is composed of oceans and seas, increasing microplastics pollution load in the marine ecosystems is bound to severely affect most of the life on the planet. Toxicological effects of microplastics on marine ecosystems are increasingly being revealed from studies (Bajt 2021). Ecotoxicity and genotoxicity associated with microplastics pollution are of serious nature and apply to ecological risk assessment of microplastics (Jiang et al. 2019). Microplastics can even induce antibiotic resistance with grave consequences for the evolution of aquatic bacteria as well as for human health (Arias-Andres et al. 2018).

More than 300 million tonnes of plastic wastes add to the planet's pollution burden each year (Prasath and Poon 2018), most of which, obviously, end up in marine waters where they trigger multiple problems to aquatic life and, through nature's food chain, even to terrestrial life. The released products are also influenced by abiotic and biotic factors. Aggregation is one of the most important environmental behaviors of the microplastics determining their mobility, distribution, and bioavailability (Wang et al. 2021).

Considerable impacts of microplastics on the environment and biota and public health, as already discussed in other chapters, are of significant concern, calling us to adopt appropriate and workable measures for the removal and remediation of microplastics pollution.

5.1 MICROORGANISMS INVOLVED IN MICROPLASTIC DEGRADATION

Both biological and non-biological methods are being employed for controlling the burgeoning problem of microplastics. Since polymer degradation is closely related with extracellular enzymes, research intervention on microplastics management is focused on the role of microorganisms. Various microbial characteristics, in addition to various environmental factors, affect microplastic biodegradation (Yuan et al. 2020). Microorganisms prevailing in waters and soils, such as bacteria, bacterial consortia, protozoa, microalgae (diatoms, flagellates, and protists), fungi, and biofilms, can effectively biodegrade microplastics. A variety of microorganisms playing a role as degraders, in essence, are widely distributed in various components of the biosphere. The biodegradation processes involve multiple biochemical reactions. Microplastics are degraded into useful products, CO_2 and H_2O, only partially under

DOI: 10.1201/9781003312086-5

laboratory conditions (Du et al. 2021). However, only a few microorganisms participate in microplastics biodegradation. Diversity of the polymer-degrading microorganisms, as it also occurs in the case of nature's biodiversity, prevails in accordance
with environmental conditions, such as saline waters, freshwaters, sludge, sediments,
and soils. Reduction of microplastics requires development of functional microbial
agents.

Various types of plastics and microplastics largely studied for their degradation/
biodegradation processes (as also mentioned in Table 5.1) include polyethylene (PE),
low-density polyethylene (LDPE), linear low-density polyethylene (LLDPE), polyethylene terephthalate (PET), polycaprolactone (PCL), polyhydroxybutyrate/poly(3-hy-
droxybutyrate-co-3-hydroxyvalarate) (PHB/PHBV), polybutylene succinate (PBS),
bisphenol A (BPA), polyvinyl alcohol (PVA), polyester (PES), polyester polyurethane
(PUR), and polystyrene (PS). Plastic wastes increasingly becoming intolerable nui-
sances for environment and life emanate from various types of plastic products being
manufactured for multiple uses in human systems.

On the basis of the degradation pathways, the synthetic plastics are categorized
into (Mohanan et al. 2020): (i) plastics having a carbon-carbon backbone, and (ii)
plastics having heteroatoms in the main chain. The plastics with carbon-carbon backbone include PE, PP, PS, and polyvinyl chloride (PVC). PUR and PET, on the other
hand, fall in the group of the plastics with heteroatoms. Petrochemical hydrocarbons are the sources from which the synthetic polymers most commonly used in our
times are derived (Geyser 2017). However, several types of biodegradable polyesters—for example, polyhydroxyalkanoate (PHA) and polylactic acid (PLA)—have
also been manufactured for use as an alternative to petroleum-based plastics. Of
these, PHAs are especially remarkable because of their high degradability in different environments, their renewability, biocompatibility, and emission of non-toxic as
well as non-polluting products upon their biodegradation (Bhatia et al. 2019a, 2019b).
Chemical structures of some major polymers that are interesting from a degradation
point of view are illustrated in Figure 5.1.

(Petro-)polymers, notably PE, PS, PU, PP, PET, and PVC, are recalcitrant to natural biodegradation pathways (Mohanan et al. 2020). Some microorganisms, as shown
in Table 5.1, have been isolated while playing a degradation role under *in situ* conditions and characterized under lab conditions. Plastic degradation, however, is dependent on a number of factors, especially on molecular weight, chemical composition,
and the degree of crystallinity.

Polymers are large molecules comprising both regular crystals constituting a
crystalline region and irregular groups constituting an amorphous region. The
amorphous region provides polymers with quite a high degree of flexibility. The
polymers with a high degree of crystallinity demonstrate lower rates of microbial
degradation (Mohanan et al. 2020). For example, PE with 95% and PET-based
plastics with 30–50% crystallinity show very high degree of rigidity to their
biodegradation.

Impacts of microplastics are more critical than those of macroplastics (Prasath
and Poon 2018). While macroplastics litter can be controlled by physical/mechani-
cal means, microplastics remediation needs specific strategies to be formulated and
implemented.

PET

PU

PE

PVC

PP

PS

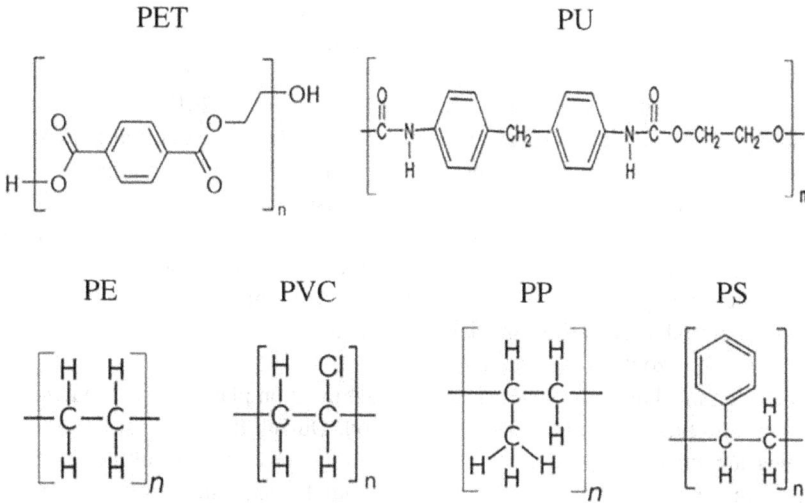

FIGURE 5.1 Structure of some important synthetic polymers of commercial use.

Note: PET—polyethylene terephthalate; PU—polyurethane; PE—polyethylene; PVC—polyvinyl chloride; PP—polypropylene; PS—polystyrene.

Microbes can adopt any environment and carry the ability to degrade different substances, including microplastics, without harming the environment. A few bacteria and fungi occurring in deep sea sediments, benthic zones of marine environments, Arctic soil, and the deep sea environment that contribute to plastic/microplastic degradation have been reviewed by Urbanek et al. (2018). As many as 90 microbial genera serving to degrade plastics were reported earlier by Chee et al. (2010). Microbial community composition varies in accordance with the specificities of habitat they prevail in. Reports on the degradation of polycaprolactone (PCL), a biodegradable polymer, suggest the involvement of *Pseudomonas* and *Rhodococcus* spp. As Urbanek (2017) observed, the PCL degradation process involving two bacteria *Pseudomonas* spp. and *Rhodococcus* spp., along with two fungal strains, results in 53% degradation (w/w) of the PCL film in 30 days of incubation.

Wang et al. (2021) found some interesting features of the bacterial and fungal communities established on microplastics and plastic sheets: (i) the communities of bacteria established were different; (ii) alpha diversity of the fungal community was not strikingly different; (iii) Ascomycota, Basidiomycota, Blastocladiomycota, and Mucoromycota were the dominant members of the fungal community; and (iv) bacteria-fungus interactions on microplastics were more complex on polyethylene (PE) and polypropylene (PP) sheets than on microplastics.

5.1.1 BACTERIAL-MEDIATED MICROPLASTIC DEGRADATION

Bacteria are adapted to survive under various marine environmental conditions. They can survive even under pretty inhospitable conditions, such as in a rock surface, in glacial and sea ice, snow, lakes with permanent ice cover, permafrost soils, cloud droplets, etc. (Cameron et al. 2012; Yadav et al. 2017; Urbanek et al. 2018), where numerous other life forms cannot survive. Abundance of bacteria in an ocean ecosystem can be imagined from the fact that hundreds of millions of bacterial cells are found per gram of wet ocean sediment (Harrison et al. 2011). The floating and the sunken plastics and microplastics are also not free from bacterial colonization. Bacterial colonization on plastics, in fact, begins immediately in marine ecosystems along with bacterial assemblages covering the plastic surfaces (Urbanek 2018). This process, known as attachment, is part of the bacterial strategy to prosper around plastic wastes, developing microfilm to carry on further degradation processes. The subsequent developments following attachment in the marine environment include formation of stable consortia, nutrient accumulation, protection against toxic substances, and horizontal gene exchange.

The exact mechanisms of the bacterial attachment on plastics in the marine environment, however, are so far poorly understood. During this process, the microbial colonies are engaged in catalyzing the metabolic processes resulting in desorption, adsorption, and breakdown of the compounds associated with microplastics. Fragmentation of plastic wastes in the aquatic environment might also take place during the processes following attachment and subsequent bacterial metabolic activities.

In the process of biodegradation, the microorganisms of the biofouling community alter or transform the structure of the chemicals through the enzymes they release and metabolic activities they are involved in. The transformed chemicals get introduced into the habitat. Larger parts of the plastic wastes which, upon entering into a water body, are fragmented into tiny parts attaining the sizes of microplastics (50 μm–5 mm) and nanoplastics (<100 nm). This happens primarily due to abiotic factors, like mechanical degradation, weathering, photodegradation, oxidation, hydrolysis, etc. Biotic factors, such as microorganisms and even invertebrates and chemicals released by aquatic plants, may also contribute to the fragmentation of the plastic wastes. In addition, a mix of several abiotic factors, including the characteristics of plastic waste (for example, structure, topography, and material properties) and of the surface inviting attachment (for example, surface hydrophobicity, surface free energy, and surface electrostatic interactions) help contribute to influence the process of biofilm development. The environmental factors influencing the process are wind speed (increasing turbulence in a water body), light intensity, temperature, water potential, salinity, and oxygen level. It is owing to these operating factors in an aquatic ecosystem that the biomass of the fouling community would not always be unaffected.

Degradation of a variety of microplastics by a variety of bacteria goes on following the previously mentioned developments and tends to intensify depending on a mix of favorable factors. The microplastic-degrading bacteria release hydrolytic extracellular enzymes, for example, lipase, keratinase, CMCase, xylanase, and chitinase (Chandra et al. 2020a), into their surroundings, which play a dramatic role in the process of microplastic degradation. Polyurethane (PUR/PU), a common polymer composed of organic units joined by urethane links, depolymerizes into ester and urethane bonds on account of the hydrolytic properties of the enzymes, namely protease, esterase, and urease. Medical polyesters are liable to be degraded by urease and papain that carry proteolytic properties.

5.1.1.1 Bacterial Strains capable of Degrading Microplastics

A number of bacteria playing a role in the processes leading to microplastics degradation have been isolated and identified. Rates/extent of microbial degradation, however, have been recorded as limited (Debroas et al. 2017; Urbanek et al. 2018; Raddadi and Fava 2019). While microplastics biodegradation would only help reduce marine plastic wastes *in situ* to a smaller extent, a few strains of bacteria can play a considerably effective role towards microplastics remediation in the marine environment. Accelerated biodegradation by using identified strains of bacteria can be realized under laboratory conditions.

Yoshida et al. (2016) screened natural microbial communities exposed to poly(ethylene terephthalate) (PET) contaminated environmental samples and isolated *Ideonella sakaiensis* 201-F6 that had the ability to use the polymer as its carbon and energy source. When grown on PET, the isolated strain was found, producing two enzymes with the ability to hydrolyze PET.

Several bacterial strains capable of degrading microplastics have been reported by Chandra et al. (2020a). These include *Bacillus* sp. BCBT21, *Bacillus subtilis*, *B. cereus*, *B. vallismortis* bt-dsce 01, *B. amyloliquefaciens* BSM-1, *B. amyloliquefaciens* BSM-2, *Brevibacccillus borstelensis*, *Pseudomonas putida*, *P. protegens* bt-dsce02, *Paenibacillus* spp.bt-dsce04, and *Stenotrophomonas* spp. Bt-dsce03. Urbanek et al. (2018) isolated several microbes thriving in the cold environment of the Arctic region (Spitsbergen). Some of the isolated bacterial strains have been experimentally brought into use for biodegradation of various kinds of plastics. These are *Pseudomonas* spp. (nine strains), *P. frederiksbergensis* (two strains), and *P. mandelii*.

Five polycaprolactone (PCL)-degrading strains of bacteria belonging to genera *Pseudomonas*, *Alcanivorax*, and *Tenacibaculum* were isolated from deep ocean waters. The two of the isolates, RCL01 and TCL04, were found adapted to lower temperatures (4°C) as well as to high hydrostatic pressure (Sekiguchi et al. 2011).

Bacterial isolates *Bacillus cereus* and *Bacillus gottheilii* have been found with the capability to degrade UV-treated microplastics (Prasath and Poon 2018). With a high degree of resistance to degradation, the plastics require specific pretreatment, such as photooxidation or hydrolysis or enzymatic degradation by microorganisms. The process of biodegradation begins by means of the formation of a biofilm and secretion of extracellular enzymes playing a very crucial role in polymer degradation.

Microplastic biodegradation processes and involved biochemical reactions of microorganisms, for the sake of applicable methodological issues, are almost invariably tested under laboratory conditions. The experimental conditions, however, often vary from those in the field (Bajt 2021). For instance, microplastic mixture concentration; exposure time of UV radiation; influence of water waves, currents, and other environmental factors; health and nutrition, etc. prevailing under natural conditions and under laboratory conditions would be quite different.

Many organisms capable of degrading polymers in natural environmental conditions have been described after their isolation from diverse sources. In a few cases, the enzymes produced by the microorganisms were cloned and sequenced. Isolated from diverse environments, many bacterial strains and fungi have been used in experiments for their bio-degradability behavior under laboratory conditions. Some of the bacterial strains playing a role in the degradation of specific plastics/microplastics are presented in Table 5.1.

TABLE 5.1
Various bacterial strains with ability to degrade different types of microplastics.

Type of plastics/ microplastics	Bacterial strains capable to degrade	Source	References
Polyethylene (PE)	*Pseudomonas* spp.	Bay of Bengal	Sudhakar et al. (2007a)
	Bacollus cereus, Bacillus sphericus spp.	Marine ecosystem	Sudhakar et al. (2008)
	Enterobacter asburiae YT1, *Bacillus* spp. YP1	Gut of the Indian mealmoth larvae (*Plodia interpunctella*)	Yang et al. (2014)
	Bacillus spp., *Paenibacillus* spp.	Municipal landfill sediment	Park and Kim (2019)
	Bacillus gottheilii, B. cereus	Mangrove ecosystem	Auta et al. (2017)
Low-density polyethylene (LDPE)	*Brevibacillus borstelensis* strain 707	Soil	Hadad et al. (2005)
	Microbulbifer hydrolyticus IRE-31	Lignin-rich pulp wastes in marine environment	Li et al. (2020)
Low molecular weight polyethylene (LMWPE)	*Pseudomonas* spp. E4	Beach soil contaminated extensively by crude oil	Yoon et al. (2012)
Polystyrene (PS)	*Rhodococcus ruber* C208	Biofilm producing strain	Mor and Sivan (2008)
	Bacillus gottheilii	Mangrove ecosystem	Auta et al. (2017)
Polypropylene (PP)	*Rhodococcus* spp., *Bacillus* spp., *Bacillus gottheilii*	Mangrove sediments	Auta et al. (2018)
	Citrobacter spp., *Enterobacter* spp.	Guts of the two larvae, *Tenebrio molitor* and *Zophobas atratus*	Yang et al. (2021)
Polycaprolactone (PCL)	*Pseudomonas* spp.	Deep seawater in Tottori Prefecture and Toyama bay offshore	Sekiguchi et al. (2009)
	Shewanella, Moritella spp., *Psychrobacter* spp., *Pseudomonas* spp.	Deep sea sediment, the Kurile and Japan Trenches	Sekiguchi et al. (2010)
Polycaprolactone (PCL), bioplastic bags made from potato and corn starch	*Pseudomonas* spp. (strains 2B, 5D, 23B, 28E, 31C, 42C, 42D, 42E, 52D), *P. frederiksbergensis* (strains 4A, 31A), *P. mandelii* (strain 33C)	Arctic Region, Spitsbergen soil	Urbanek et al. (2017)
Polyurethane-diol (PUR-diol)	*pseudomonas aeruginosa*	Soil	Mukherjee et al. (2011)
Polyethylene glycols (PEGs)	*Pseudomonas stutzeri* JA1001	River water	Obradors and Aguilar (1991)

Type of plastics/ microplastics	Bacterial strains capable to degrade	Source	References
Polyethylene terephthalate (PET)	*Ideonella sakaiensis* 201-F6	Natural microbial communities in the environment	Yoshida et al. (2016)
Polyethylene (PE), polyethylene terephthalate (PET), polypropylene (PP), and polystyrene (PS)	*Bacillus gottheilii*	Mangrove ecosystem	Auta et al. (2017)
PE, PS, PET	*Bacillus cereus*	Mangrove ecosystems	Auta et al. (2017)
PCL (monofilament fibers), polyhydroxybutyrate/ poly(3-hydroxybutyrate-co-3-hydroxyvalarate) (PHB/PHBV), polybutylene succinate (PBS)	*Pseudomonas* spp., *Alcanivorax* spp., *Tenacibaculum* spp.	Deep seawater	Sekiguchi et al. (2011)
Bisphenol A (BPA)	*Pseudomonas* spp.	Marine ecosystem, Bay of Bengal	Artham and Doble et al. (2009)
Polyvinyl alcohol (PVA), Linear low-density polyethylene (LLDPE)	*Vibrio alginolyticus, Vibrio parahemolyticus,*	Benthic zones of the marine environments	Raghul et al. (2014)
Nylon 6 and 66	*Bacillus cereus, Bacillus sphericus, Vibrio furnisii,* and *Brevundimonas vesicularis*	Marine environment	Sudhakar et al. (2007b)

Microbial communities inhabiting diverse environmental conditions, often pretty unfavorable too, reveal many unique characteristics. The growing plastic pollution menace in the marine ecosystems might allure benthic organisms to use plastics as a substratum. Dissolved organic carbon released from plastics in seawater, obviously, would help the heterotrophic organisms, including bacteria, to play their role. Adaptation to new carbon as a source of energy may lead to evolve new features of the newly established organisms, especially in producing and secreting enzymes active even under unfavorable conditions, such as in the cold regions that most of the marine ecosystems on the planet lie in. These new features provide ample biotechnological development opportunities for us to solve the grave planet problems, including plastic pollution and the emerging microplastic menace. Microorganisms adapted to relatively unfavorable conditions might also be employed in sanitary landfill areas. The frequently quoted bacteria being employed for biodegradation, according to Pathak and Navneet (2017), belong to *Arthrobacter, Corynebacterium, Micrococcus, Pseudomonas, Rhodococcus,* and *Streptomyces.*

As many as 35 bacteria isolates with morphological distinctiveness prevailing in the cryoconite holes at glaciers in Spitsbergen were isolated by Singh et al. (2014), which belonged to eight genera, namely, *Pseudomonas, Polaromonas, Micrococcus, Subtercola, Agreia, Leifsonia, Cryobacterium*, and *Flavobacterium*. Twelve strains carried the ability to produce lipase. *Micrococcus* spp. MLB-41 expressed high amylase activity and *Cryobacterium* spp. MLB-32 showed amylase, protease, and lipase activities. This study attempts to unfold the potential applications of the microorganisms in the biodegradation of microplastics in extremely cold ecosystems. The lipase, a lipid-hydrolyzing enzyme, is also able to hydrolyze ester bonds in some polyesters. The lipase enzyme secreting microbial strains of the genera *Pseudomonas, Pseudoalteromonas, Colwellia, Marinomonas*, and *Shewanella* were earlier isolated by Yu et al. (2009) from the ice in the Arctic sea. Such psychrophilic and/or psychrotolerant microbial strains with capability to produce extracellular lipase even at 0°C (Yu et al. 2009) and capable of hydrolyzing polyesters like polycaprolactone (Pathak and Navneet 2017) are of potential significance for microplastic degradation in the cold marine ecosystems.

Marine microflora are reported to possess greater ability for the production of enzymes as well as protein active compounds. *Bacillus* spp., *Pseudomonas* spp., *Achromobacter* spp., *Arthrobacter* spp., and *Alcaligenes* spp. have been screened for lipase production. Bacterial strains, namely, *Bacillus subtilis, B. nealsonii* S2MT, *Pseudomonas alcaligenes, P. aeruginosa, P. fluorescens* BJ-10, and *P. fragi* produce lipases in higher amounts (Chandra et al. 2020b).

Apart from lipases, there are many other enzymes with capacity of degradation that many microbial strains produce. These, as reported by Pathak and Navneet (2017), are esterases, cutinases, ureases, dehydratases, proteinases, and depolymerases. Among the depolymerases are PHA depolymerases, PHB depolymerases, PLA depolymerases, and PCL depolymerases. Proteinases include proteinase K acting against PLA. Earlier, Huston et al. (2000) revealed that the highest amounts of proteases are secreted by the bacteria that exist at −1°C. This finding unfolds the potential capability of polar bacteria to produce enzymes with the properties of microplastics biodegradation. Microplastics to originate from the fragmentation of petrochemical plastics, such as polyethylene terephthalate (PET), as found in a study, are likely to undergo biodegradation by PETase, an enzyme secreted by *Ideonella sakaiensis* (Austin 2018).

Auta et al. (2022) isolated eight bacteria from the microplastics in the mangrove soil and used indigenous microbial consortium for *in situ* bioremediation of the mangrove soil polluted with PET and PS microplastics. The microbial consortium proved to be of adequate capability for degrading the microplastics.

A number of the strains of *Pseudomonas* and *Bacillus* carry properties that partially biodegrade a variety of petro-plastics, like PE, PP, PS, PUR, PET, and PVC (Mohanan et al. 2020). Thermal sensitivity and UV-radiation-absorbing capacity are the two main plastic characteristics that determine the rates of their biodegradation. Supplementation of some additives that affects these plastic features tends to increase the microplastics biodegradation process.

5.1.2 Fungal-Mediated Microplastic Degradation

A variety of fungi, like bacteria, have potency to utilize plastics/microplastics. Fungi play their degradation role through promoting formation of different chemical bonds—for example, carbonyl, carboxyl, and ester functional groups—in microplastics, thus declining their hydrophobicity, a necessary condition contributing to enhance biochemical reactions.

Fungi, not bacteria, are predominantly responsible for degrading polyester polyurethane (Barratt et al. 2003). A few types of fungi, such as *Fusarium oxysporum* and *F. soloni*, have been examined for their action upon PET (Yoshida et al. 2016; Iram et al. 2019). Fungi species, namely *Rhizopus delemar, R. arrhizus, Achromobacter* spp., and *Candida cylindracea* have been found producing lipases and esterases (Iram et al. 2019), the enzymes that aid in microplastics degradation.

Urbanek et al. (2018) isolated many fungal species from the Arctic region (Spitsbergen) and two of them, namely *Clonostachys rosea* strain 16G and *Trichoderma* spp. strain 16H, were found effective for biodegradation. Chandra et al. (2020b) reported that *Penicillium* spp., *Fusarium* spp., and *Aspergillus* spp. were screened for their lipase production and *P. expansum, P. chysogenum, A. niger*, and *Trichoderma* produce lipases in pretty high amounts.

Aspergillus spp. have been isolated from marine coastal area in the Gulf of Mannar in India and designated as *Aspergillus tubingensis* VRKPT1 and *Aspergillus flavus* VRKPT2. These fungal strains have been tested efficient in high-density polyethylene biodegradation (Sangeetha Devi et al. 2015). Different fungal species occurring in diverse soils have been found playing a remarkable role in the degradation process of polyester polyurethane. Soil type, however, is not a factor determining the physical process of degradation. *Geomycene pannorum* and *Phoma* spp. have been discovered as dominant fungal species recovered from the polyurethane sheets contaminating the acidic and the neutral soils, respectively, and polyurethane is highly prone to biodegradation by these fungi (Cosgrove et al. 2007).

5.1.2.1 Examples

Fungal bioremediation of microplastics, an eco-friendly and efficient method, has been used extensively by various researchers. Some examples of the various fungal strains collected from diverse sources and exhibiting biodegradation potential against specific microplastic types are presented in Table 5.2.

5.1.2.2 Fungi Associated with Polyethylene Microplastics Degradation

Polyethylene (PE) products are extensively used across the world. This material has become almost inevitable due to its attribute of "convenience" in daily life. Some fungal species utilizing the microplastics as source of carbon and energy may be of critical significance in countering the threat posed by the accumulation of microplastics emerging from PE products.

Bioremediation of microplastic wastes using fungi is an environment-friendly and efficient method (Sánchez 2020). Unlike that of bacteria, the fungal species' capability of biofilm formation on PE helps progressively reduce hydrophobicity of the surface (Gilan et al. 2004).

Naturally occurring fungus *Zalerion maritimum* is capable of utilizing PE, decreasing its mass as well as pellet sizes. Microplastics requiring minimum nutrients may undergo biodegradation using this fungus (Paço et al. 2017).

A fungal strain of the species *Aspergillus flavus* isolated from gut content of *Galleria mellonella* and designated as PEDX3 has been found to degrade HDPE

TABLE 5.2

Some fungal species with the ability of biodegradation of different polymers.

Microplastic types	Fungal strain	Source	References
PE, HDPE	*Aspergillus tubingensis* VRKPT1, *A. flavus* VRKPT2	Coastal area, Gulf of Mannar, India	Sangeetha Devi et al. (2015)
PE, HDPE	*Aspergillus flavus* PEDX3	Gut contents of wax moth *Galleria mellonella*	Sánchez (2020)
PET, PS, PE	*Penicillium raperi*, *P. glaucoroseum*, *Aspergillus flavus*	Soil, activated sludge, leaves, and worms' excreta	Taghavi et al. (2020)
PUR	*Geomyces pannorum*, *Phoma* spp.	Sandy loam soils/acidic and neutral soils	Cosgrove et al. (2007)
PE (500 h with UV)	*Penicillium simplicissimum* YK	Soils of coastal area	Yamada-Onodera (2001)
LDPE	*Aspergillus* spp., *A. versicolor*	Seawater	Pramila and Vijaya Ramesh (2011)
LDPE	*Chamaeleomyces viridis* JAKA1	Soil from solid waste dumping site	Gajendiran et al. (2016a)
LDPE	*Aspergillus niger*, *A. flavus*, *A. terreus*, *A. fumigatus*, *Penicillium* spp.	Red Sea water	Alshehrei (2017)
LDPE	*Aspergillus clavatus* JASK1	Landfill soil	Gajendiran et al. (2016b)
LDPE	*Aspergillus versicolor* JASS1	Municipal landfill solid waste soil	Gajendiran et al. (2017)
LDPE	*Penicillium pinophilum* ATCC11797	Strain center (woody plants)	Volke-Sepulveda (2001)
PE	*Zalerion maritimum*	Marine ecosystem	Paço et al. (2017)
PCL (bioplastic bags made from corn and potato starch)	*Clonostachys rosea* 16G, *Trichoderma* spp. 16H	Spitsbergen soil, Arctic region	Urbanek et al. (2017)
PCL	*Pseudozyma jejuensis*	Orange leaves in Jeju island in South Korea	Seo et al. (2007)
PUR	*Pestalotiopsis microspora*	Screening of endophytic fungi	Russell et al. (2011)

Note: HDPE—high-density polyethylene; LDPE—low-density polyethylene; PCL—polycaprolactone; PE—polyethylene; PET—polyethylene terephthalate; PS—polystyrene; PUR—polyurethane.

into lower molecular weight microplastic particles in a four-week incubation period. The appearance of the carbonyl and ether groups of the microplastic particles also confirmed PE degradation by the fungal strain PEDX3. The fungal strain produces two PE degrading enzymes, laccase-like multicopper oxidases (LMCOs), which may provide a promising application for PE microplastics remediation (Zhang et al. 2020).

Microplastics lead to alteration in the composition of fungal communities in the aquatic environment. Kettner et al. (2017) discovered fungal diversity prospering on the particles of PE and PS incubated in the waters of the differing ecosystems: the Baltic Sea, the river Warnow, and in a wastewater treatment plant. The interesting findings they presented are (i) 347 taxonomic units belonging to 81 fungal taxa prosper on both the polymers; (ii) communities associated with microplastics are distinctive from those associated with wood substrate in the surrounding water; (iii) there is a significant difference in accordance with sampling locations; and (iv) saprophytic and parasitic fungi belonging to Ascomycota, Chytridiomycota, and Cryptomycota as the dominant fungal assemblages thrive in microplastic biofilms. Important fungal strains active in the PE microplastic biodegradation process have been listed in Table 5.2.

5.2 MECHANISMS INVOLVED IN THE BIODEGRADATION OF MICROPLASTICS

All heterotrophic microorganisms are the organisms that play a role in decomposing the complex organic matter into the simpler organic/inorganic forms leading to mineralization. In other words, the decomposers (also called biodegraders or reducers) are the microorganisms that are capable of reducing the organic matter into the constituents it is originally made up of. This transformation, mediated by the microorganisms, involves biochemical reactions which are driven by the enzymes released by the microorganisms. Biosynthesis of the proteins functioning as enzymes is under genetic control. The participating microorganisms utilize carbon and energy for their metabolism, growth, and functions. The biodegradation is an energy-releasing process.

The enzymes in action on the polymer surface are external or exo-enzymes. The external enzymes may have wide functionality, for instance, oxidative as well as hydrolytic. The enzyme action initially breaks down larger and complex polymers into their smaller and simpler units. The microbial action on polymers can also be referred to as depolymerization. The transformative microbial enzyme action on polymers registers reduction in its molecular weight, changes in surface properties, and decline in mechanical strength. The amorphous region in the plastics generally is more prone to degradation by hydrolysis than the crystalline one because water is more easily penetrable into the former region than into the latter.

The biodegradation of a C-based polymer occurring in aerobic conditions can be represented through the following equation:

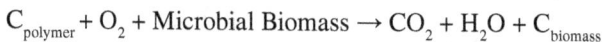

$$C_{polymer} + O_2 + \text{Microbial Biomass} \rightarrow CO_2 + H_2O + C_{biomass}$$

O_2 in the degradation process acts as a terminal electron acceptor. Thus, when the biodegradation sets in under aerobic conditions, water and carbon dioxide are produced. When anaerobic conditions prevail in the environment in which the biodegradation process occurs, the degradation products are CO_2, H_2O, and CH_4. Polymer biodegradation under aerobic conditions is more efficient than that under anaerobic conditions. A greater amount of energy is also released under aerobic conditions, which is an attribute for the microbial community to thrive more profusely.

Assimilation of carbon the polymer is composed of ($C_{polymer}$) by the microorganisms increases carbon in the biomass ($C_{biomass}$) at the cost of CO_2 and H_2O lost into the environment. $C_{biomass}$ keeps dynamically undergoing mineralization through bio-chemical processes. How much $C_{polymer}$ translates into $C_{biomass}$ and other products would depend on the efficiency of biological degradation of the polymer which depends on many operating environmental factors, polymer characteristics, and type of participating organisms.

A polymer is not converted directly into end products. Intermediary products are most likely to exist. For example, a polymer would first break into oligomers, dimers, and monomers that would undergo the same fate. Moreover, only a small proportion of carbon ($C_{polymer}$) is converted into biomass ($C_{biomass}$), a proportion of which remains as a residual carbon ($C_{residues}$). Further, polymer disintegration by microbial actions is an energy-releasing process in which energy is released in the form of heat. Therefore, the following equations suggest still more comprehensive description of the polymer biodegradation under aerobic conditions:

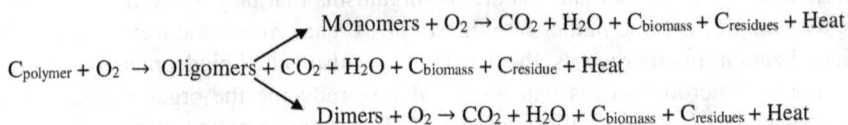

$$\text{Monomers} + O_2 \rightarrow CO_2 + H_2O + C_{biomass} + C_{residues} + \text{Heat}$$

$$C_{polymer} + O_2 \rightarrow \text{Oligomers} + CO_2 + H_2O + C_{biomass} + C_{residue} + \text{Heat}$$

$$\text{Dimers} + O_2 \rightarrow CO_2 + H_2O + C_{biomass} + C_{residues} + \text{Heat}$$

$C_{polymer}$ and newly formed oligomers get converted into $C_{biomass}$; however $C_{biomass}$ gets converted to CO_2 under a different kinetics scheme (Glaser 2019). These chemical equations also reveal how the physical and chemical changes do take place in the process of polymer biodegradation.

When biological agents of degradation are not capable of penetrating into the bulk layer of the polymer, their degradation role remains confined to the surface of the material and if they succeed in penetrating the surface layer, they would participate in bulk degradation. The former type of biodegradation may be marked by the presence of spherulites.

Aerobic biodegradation of the polymers, which is generally emphasized while dealing with microplastic pollution, may not always be the case. In various environments anaerobic conditions do prevail which avert the occurrence of aerobic respiration. For example, the buried plastic material would first undergo aerobic degradation. When the aerobic microbes consume the entire amount of oxygen, the anaerobic degradation process is switched on. Alternate electron acceptors, such as nitrate, sulfate, or methanogenic conditions, enable the initiation of anaerobic biodegradation (Glaser 2019) and if the oxygen again becomes available at the biodegradation site, anaerobic biodegradation ceases to take place. Anaerobic degradation processes are relatively less efficient and slower than the aerobic ones.

Degradation of a material in the environment is not attributable to a single mechanism but to a combination of mechanisms. In nature, in the process of organic matter degradation, biotic and abiotic factors act synergistically. Degradation, in fact, includes disintegration and mineralization. The disintegration is attributable to photodegradation, mechanical degradation, hydrolytic degradation, oxidation, weathering, and the action of exo-enzymes of microbial origin. Disintegration, the initial phase of degradation, leads to the formation of microplastics. Mineralization, the final phase of degradation, occurs when the participating microbes metabolize

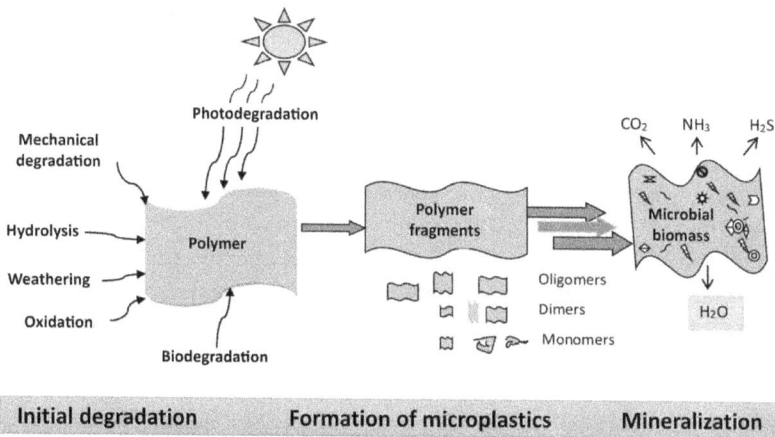

FIGURE 5.2 Outline diagram of plastic degradation.

the microplastics, converting them into inherent digestion/metabolic products. The mechanism has been illustrated in Figure 5.2.

Assimilation involves integration of the atoms from the polymer fragments into microbial cells. During the biodegradation/depolymerization process, the participating microorganisms derive C, H, O, N, P, S, and other elements and molecules which all participate in cellular metabolic activities. The elements absorbed and assimilated by the microbes are subject to metabolic (catabolic and anabolic) activities vital for growth, reproduction, and decomposition activities. The molecules for which microbial cell membranes are impervious undergo bio-transformation reactions through which the products worth assimilation by the microbes are produced.

According to a new analysis, the polymer biodegradation is referred to as a complex process involving three stages: biodeterioration, biofragmentation, and assimilation (Glaser 2019; Iram et al. 2019). The first stage of polymer biodegradation taking place on the plastic surface is largely attributable to abiotic factors. Resulting in the alteration of mechanical and physicochemical characteristics of the material (Lucas et al. 2008), this phase is known as biodeterioration. Abiotic degradation, in fact, is an essential step in the degradation process of synthetic polymers, such as polycarboxylates, PET, PLA and their copolymers, poly(alpha-glutamic acids), and poly(dimethyl) siloxanes, or silicones (Shah et al. 2008). The pollutants, upon being adsorbed, may assist the polymer colonization by microbial species, namely, bacteria, fungi, protozoa, microalgae, as the possible participants in the process (Glaser 2019). Microorganisms, in addition to the environmental factors, assist in chopping down biodegradable polymers into small fragments (Iram et al. 2019).

The subsequent phase of biofragmentation (also known as depolymerization) involves cleavage of molecular bonds of the polymers as a result of the exo-enzymes secreted by microorganisms, such as lipase, proteinase K, hydrogenase, esterase, polyurethanase, cutinase, monoxygenase, etc., yielding low molecular weight polymers including oligomers, dimers, and monomers. Various enzymes the microorganisms secrete have the potential to degrade various synthetic polymers (Iram et al. 2019). However, enzyme proteinase K (also known as protease K or endopeptidase K), a broad-spectrum serine protease secreted by the fungal spp. *Engyodontium album* (earlier known as *Tritirachium album*), has been reported to be a potent degrader of PLA. Many strains of *Amylocolatopsis* and *Saccharothrix* also have potential to biodegrade PLA (Ghosh et al. 2013; Iram et al. 2019).

In the assimilation stage, monomers, dimers, and oligomers, upon being recognized by cell membrane receptors, are absorbed into microbial cells. The non-recognizable ones are subject to biotransformation for enabling them to make entry into microbial cells via cellular membrane (Lucas et al. 2008). Various molecules having absorbed into microbial cells undergo biochemical pathways towards the synthesis of packaging vesicles, fresh biomass, various primary and secondary metabolites, and energy-yielding molecule ATP. Some of the metabolites (for example, acids, aldehydes, antibiotics, terpenes, etc.) get released from the microbial cells into the surrounding environment. Mineralization, a dynamic process releasing basic inorganic constituents into environment, goes on through aerobic or anaerobic pathways.

Aerobic biodegradation, while the participating microorganisms perform aerobic respiration, involves use of oxygen as an electron acceptor. The process results in the breakdown of polymers—the oligomers, dimers, and monomers—into carbon biomass and carbon residue, releasing CO_2 and H_2O. The aerobic biodegradation can be represented by the following equation:

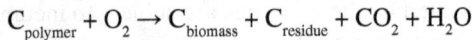

$$C_{polymer} + O_2 \rightarrow C_{biomass} + C_{residue} + CO_2 + H_2O$$

Anaerobic biodegradation involves the participation of anaerobic microorganisms in the absence of oxygen. In this process, methane, manganese, iron, sulfate, nitrate, and carbon dioxide serve as electron acceptors. The process can be represented by the following equation:

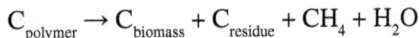

$$C_{polymer} \rightarrow C_{biomass} + C_{residue} + CH_4 + H_2O$$

A broad schematic representation of microplastic biodegradation under aerobic and anaerobic conditions is shown in Figure 5.3.

Biodegradation of a polymer to the extent of 100% is impossible because a fraction of it is incorporated into microbial biomass ($C_{biomass}$) and residues ($C_{residues}$).

5.2.1 MICROBIAL ENZYMES INVOLVED IN THE DEGRADATION OF MICROPLASTICS

Microplastics degradation processes are closely related to the reactions of enzymes produced by specific microorganisms, especially the bacteria and the fungi. More than 90 microorganisms, including fungi and bacteria, are capable of biodegrading petroleum-based plastics (Jumaah 2017) mostly *in vitro* conditions. A few enzymes playing a role in microplastics degradation have been isolated from certain microorganisms and their specific characteristics vital for polymer degradation have been explained in several studies as summarized in Tables 5.3 and 5.4.

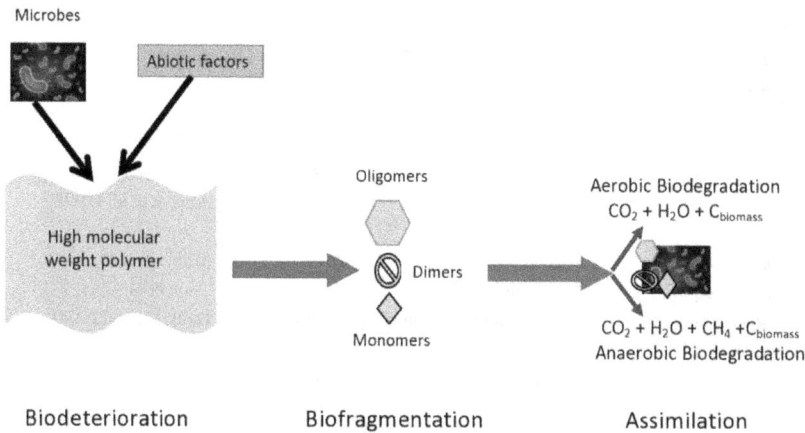

FIGURE 5.3   A schematic presentation of polymer degradation.

The participating microorganisms utilize carbon and energy for their metabolism upon the amorphous region and then on the crystalline region. Hydrolysis, the first step of the biodegradation, enhances hydrophobicity by the receptive enzyme imparting functional groups (C–O and C=O) to the polymer in the process of depolymerization. This process increases the polymer's susceptibility to the biodegrading microorganisms (Iram et al. 2019). The hydrolytic reactions are initiated by hydrolase enzymes, a class of enzymes comprising a number of enzymes, such as glycosidase, esterase, lipase, phosphatase, etc. (Roohi et al. 2017).

Acting upon large plastic polymers, the extracellular enzymes, such as hydrolases and depolymerases, reduce them into smaller molecules (Shah et al. 2008). Hydrolytic cleavage of a polymer may take place as a result of exo-attack (at the polymer chain terminus) or endo-attack (at some point along the polymer chain). Oligomers and monomers produced as a result of exo-attack are assimilated by bacteria into their cells. Reduced molecular weight as a result of endo-attack, however, is not assimilated without further degradation (Lenz 1993).

The microplastic degradation involving the action of specific enzyme is a process taking place in two steps: (i) enzyme binding with the polymer substrate, and (ii) hydrolytic cleavage of the polymer. A variety of microorganisms varying according to environmental specificities (for instance, sea, river, municipal wastewater, activated sludge, landfill soil, field soil, sediment, compost, etc.) are attracted to colonize plastic debris. Extracellular enzymes released by the microorganisms play their degradation role based on the structural constituents of a colonized polymer.

Various strains within a diversity of bacteria have been found producing enzymes identified as PUR esterase, lipase, alkane hydroxilase, alkene monooxigenase, cutinase, esterase, PUR protease, PEG dehydrogenase, serine hydrolase, aryl acylamidase, laccase, etc. Each of the enzymes breaks down the specific linkages a polymer has between its constituent chemicals. Some of the enzymes produced by certain bacterial strains and specialized in degrading particular plastic types are presented in Table 5.3.

TABLE 5.3

Enzymes produced by various bacterial strains and the plastic types they degrade.

Bacteria	Enzymes produced	Plastics	References
Brevibacillus borstelensis	Unknown	LDPE	Hadad et al. (2005)
Comamonas acidovorans TB-35	PUR esterase	PUR	Akutsu et al. (1998)
Pseudomonas spp.	Lipase	PET	Müller et al. (2005)
Pseudomonas spp. E4	Alkane hydroxylase	LMWPE	Yoon et al. (2012)
Pseudomonas putida AJ, *Ochrobactrum* TD	Alkene monooxygenase	VC/PS	Danko et al. (2004)
Pseudomonas protegens Pf-5	Lipase	PUR	Hung et al. (2016)
Pseudozyma jejuensis OL71	Cutinase	PCL	Seo et al. (2007)
Pseudomonas spp. AKS2	Esterase	PES	Tribedi and Sil (2014)
Pseudomonas protegens BC2–12	Lipase	PUR	Hung et al. (2016)
Pseudomonas protegens CHA0	Lipase	PUR	Hung et al. (2016)
Pseudomonas fluorescens A506 and Pf0–1	Lipase	PUR	Hung et al. (2016)
Pseudomonas chlororaphis	Polyurethanase, lipase	PUR	Stern and Howard (2000)
Pseudomonas aeruginosa MZA-85	Esterase	PUR	Shah et al. (2013)
Pseudomonas vesicularis PD	Esterase	PVA	Sakai et al. (1998)
Pseudomonas fluorescens	Polyurethanase-protease	PUR	Howard and Blake (1998)
Pseudomonas stutzeri JA1001	PEG dehydrogenase	PEG	Obradors and Aguilar (1991)
Pseudomonas stutzeri	Serine hydrolase	PHA	Shimao (2001)
Pseudomonas chlororaphis	Polyurethanase	PUR	Ruiz et al. (1999)
Pseudomonas aeruginosa	Esterase	PUR	Mukherjee et al. (2011)
Pseudomonas spp.	Lipase	PUR	Biffinger et al. (2015)
Pseudomonas fluorescens	Esterase	PUR	Biffinger et al. (2015)
Pseudomonas fluorescens	Esterase/protease	PUR	Vega et al. (1999)
Rhodococcus equi	Aryl acylamidase	PUR	Cited by Iram et al. (2019)
Schlegelella thermodepolymerans	Unknown	P(3HB-co-3MP)/PHA	Cited by Iram et al. (2019)
Thermobifida fusca	Hydrolase	PET	Müller et al. (2005)
Thermomonospora fusca	Unknown	PVC	Kleeberg et al. (1998)
Tremetes versicolor	Laccase	Nylon, PE	Nishida et al. (2001)

Note: LDPE—low-density polyethylene; LMWPE—low-molecular weight polyethylene; P(3HB-co-3MP)/PHA—poly(3-hydroxybutyrate-co3-mercaptopropionate)/polyhydroxyalkanoate; PCL—polycaprolactone; PE—polyethylene; PEG—polyethylene glycol; PES—polyethylene succinate; PET—polyethylene terephthalate; PHA—polyhydroxyalkanoates; PUR—polyurethane; PVA—polyvinyl alcohol; PVC—polyvinyl chloride; VC/PS—vinyl chloride/polystyrene.

TABLE 5.4
Enzymes produced by some polymer-degrading fungal species.

Fungi	Enzymes	Plastics	References
Acremonium spp.	Unknown	PHB, poly[3HB-co-(10 mol%) 3HV]	Cited by Iram et al. (2019)
Agromyces spp.	Nylon hydrolase	Nylon-6 (oligomers)	Negoro et al. (2012)
Candida antarctica	Lipase	PET	Müller et al. (2005)
Amycolaptosis spp.	Manganese peroxidase	PLA, PE	Asiah et al. (2015)
Aspergillus spp. strain ST-01	Unknown	PHB	Sanchez et al. (2000)
Aspergillus clavatus strain NKCM1003	P(3HB) depolymerase	PES	Ishii et al. (2007)
Aspergillus flavus	Glycosidase	PCL	Tokiwa et al. (2009)
Aspergillus niger	Catalase, protease	PCL	Tokiwa et al. (2009)
Cephalosporium spp.	Unknown	PHB	Matavulj and Molitoris (1992)
Fusarium spp.	Cutinase	PCL	Asiah et al. (2015)
Fusarium solani f. spp. pisi strain 77–2–3	PCL depolymerase, cutinase	PCL, cutin	Murphy et al. (1996)
Deuteromycetes (*Fungi imperfecti*) 41 strains	Unknown	PHB	Lee et al. (2005)
Penicillium funiculosum	Unknown	PHB	Tokiwa et al. (2009)
Penicillium funiculosum QM301	Unknown	PBA, PEA, PPA	Cited by Iram et al. (2019)
Penicillium spp. strain 14–3	Lipase (with broad substrate specificity)	PEA, PES, PBS	Tokiwa et al. (2009)
Penicillium spp. Strain 26–1	Unknown	PCL	Tokiwa et al. (2009)
White-rot fungus IZU-154	Manganese peroxidise	Nylon	Deguchi et al. (1998)

Note: PBA—polybutyric adipate; PBS—polybutylene succinate; PCL—polycaprolactone; PE—polyethylene; PEA—polyethylene adipate; PES—polyethylene succinate; PET—polyethylene terephthalate; PHB—polyhydroxyl butyrate/poly-3-hydroxybutyric acid; PLA—polylactic acid; PPA—polypropylene adipate; PUR—polyurethane.

The main enzymes isolated from various fungi that are crucial for the degradation of a range of plastic types have been identified as nylon hydrolase, lipases, manganese peroxidase, P(3B)depolymerase, glycosidase, catalase, protease, cutinase, PCL depolymerase, etc. (Table 5.4).

Some of the extracellular enzymes tested under laboratory conditions that contribute to polymer degradation are still unknown, and in some cases the specific enzymes produced by certain microbes have been successfully cloned and sequenced. Polymers exposed to external enzymes break down into oligomers, dimers, and monomers, and eventually, through mineralization, end up into CO_2, H_2O, N_2, and minerals. However, there could be some limitations for an enzyme to degrade a polymer to the extent of 100%. Some factors—environmental, polymer related, and microorganism specific—control the degradation process, including the extent to which the degradation process would take place.

5.3 USE OF BACTERIAL BIOFILMS IN MICROPLASTIC DEGRADATION

Plastic surfaces offer a unique habitat for a variety of microorganisms from the surrounding environment (Wang et al. 2021). Microbial communities accumulate on plastic surfaces, forming a sticky substance trapping microplastic particles in what is known as a biofilm. Chain scission leading to the generation of oligomers and their radicals takes place as a result of abiotic degradation (Gewert et al. 2015), to be followed by their biodegradation. Nevertheless, the degree of plastic degradation is largely determined by the physical features and chemical properties of the polymers.

Microbial dynamics during the process of biofilm formation is remarkable: a transition from planktonic biota to a surface-attached community. Some microorganisms thriving in planktonic biota get adsorbed on the polymer surface and release exopolysaccharides which function like an adhesive for biofilm development (Figure 5.4). Microbial cells growing within the biofilm produce the substances—the composites of extracellular polysaccharides, lipids, proteins, and DNA—termed as extracellular polymeric substances (EPS) (Glaser 2019, 2020).

The microorganisms within a biofilm exhibit the phenotypic characteristics different from their planktonic stage. The microbial cells dynamically stay in a circuit of three different development phases in response to environmental signals and under genetic control, namely, (i) cell-microplastic surface interactions, (ii) biofilm maturation, and (iii) resumption to plankton stage. Pollutants available at the polymer surface may help the biofilms to grow. However, certain chemicals not suitable for the growth of the microorganisms might cease biofilm formation. Uninterrupted nutrient supplies help in the growth of a biofilm. However, if the nutrient supply is uninterrupted, the biofilm microbial community assumes planktonic stage.

The biofilm formation and development process does not proceed in a linear fashion. It is a complex process and is subject to a number of material (plastic type)-related abiotic and biotic factors (Figure 5.5) as also explained in Table 5.5.

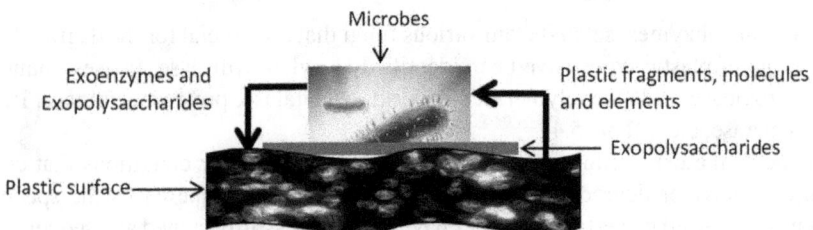

FIGURE 5.4   Schematic representation of microbial attachment to plastic surfaces.

Source: Adapted from Glaser (2019).

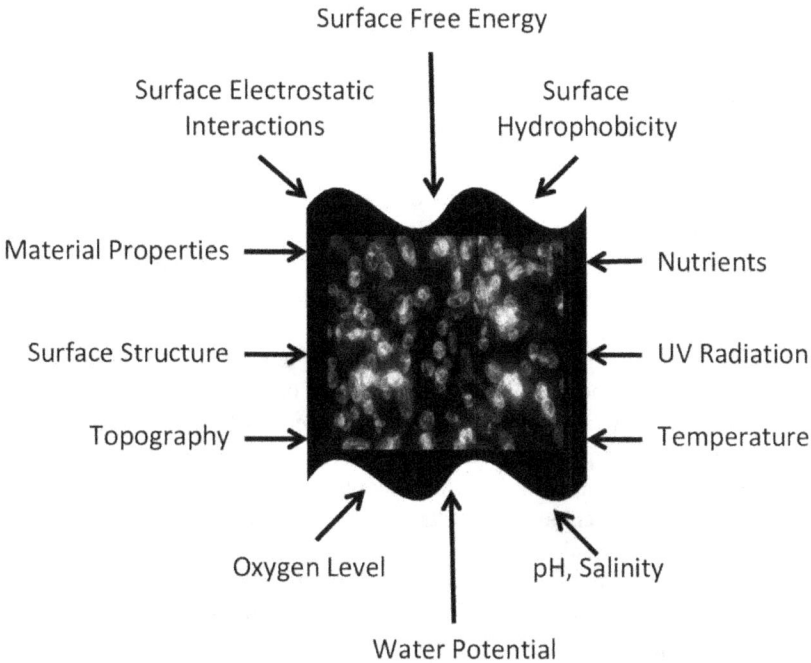

FIGURE 5.5 Various abiotic and plastic-related factors affecting biofilm formation.

Biofilm development leads to increased density of the plastic pieces due to which they sink towards the bottom of the water body. Biofouling, the bacterial adhesion and biofilm formation, involves adsorption and biomass immobilization followed by microfouling and macrofouling. Among the microorganisms, bacteria are the first to colonize, entrapping other organisms like fungi and diatoms (Selim et al. 2017). Desorption, absorption, and fragmentation of polymer chains result from the ongoing metabolic activity of the microbial biomass.

Biofilm formation facilitates the establishment of non-bacterial microorganisms (Pauli et al. 2017), for example microalgae and microscopic fungi. A biofilm represents a unique phylogenetic, functional, and ecological entity and is often referred to as a microbial assemblage, biofouling community, or periphyton (Rummel et al. 2017). As the biofilms are distinguishable from the surrounding water and are ecologically quite distinctive, they constitute what is called a plastisphere. The microalgae settling includes diatoms, flagellates, and protists. The ratios between various organisms prospering on plastic surfaces vary in accordance with environmental factors within marine ecosystems. For example, the ratios between cell counts of bacteria, diatoms, and flagellates attached to polymer pieces, as observed by Salta et al. (2013), were 640:4:1, with proportion of other organisms accounting for only 0.15%. Thus, abundance of the microorganisms perpetuating on plastics is obviously to be that of bacteria.

With increased density of the microbial community, that is biofouling community, a biofilm sinks to the floor of an aquatic ecosystem. The "plastisphere" so formed attracts invertebrates that can graze on the inhabiting organisms. Thus, the plasti-sphere is enriched by a variety of organisms. Transport of non-native or alien species is one of the consequences in this transmission process. Microorganisms inhabiting the plastisphere would now be dispersed to other parts of a water body imparting their negative influence on the physical environment as well as on the biota of an aquatic ecosystem.

Biofouling plays a critical role in governing the buoyancy of the plastic wastes. The plastic degraders assume the formation of biofilm as it is the way they increase their growth on a plastic surface. By developing a biofilm, the associated microbes become able to induce physicochemical changes on the plastic surface at a faster rate.

Hydrophobicity of polyethylene decreases during biofilm formation, resulting in the abundance of hydrophilic C–O and C=O groups on the surface (Tu et al. 2020). Hydrophobicity greatly influences bacterial attachment as well as degradation. The plastic polymers are hydrophobic in their characteristics. Therefore, bacteria, in order to facilitate their attachment, initiate hydrophobic reactions with the plastic surface. Hydrophobicity can be enhanced by starving the bacterial culture. For instance, as reported by Ghosh et al. (2019) in their review study, *Rhodococcus corallinus* when carbon starved became more hydrophobic and adhered to plastic surface more strongly and degraded polymers more efficiently. The high affinity of *Rhodococcus ruber* on polyethylene surfaces is also owing to the same reason. The possible mech-anism is that low carbon availability promotes hydrophobic interactions and biofilm establishment (Ghosh et al. 2019). Biofilms developed with interactions of other bac-teria, such as those prevailing in the soil, represent a similar scenario.

The microflora adhered to the surface introduce into the polymer non-specific bonds and several functional groups that serve to enhance degradability as well as hydrophilicity. Microflora attachment triggers processes that induce abiotic degrada-tion. Weathering of the plastic surface occurs through modification of its topography and increased roughness by means of introduced polar hydrophilic groups into the polymer, a process induced by abiotic degradation by UV light (Ghosh et al. 2019). Improved surface roughness is more appropriate for the attachment of microorgan-isms to further contribute to modifying polymer structure and composition necessary for enhanced polymer fragmentation with a higher surface-to-volume ratio which, according to Rummel et al. (2017), is an essential aspect of the plastic degradation process.

Microplastics-associated biofilms embrace the ability to influence environmen-tal processes and subsequently the fate of microplastics in the marine and coastal environments. Tu et al. (2020), in their study on biofilm formation and impacts on PE microplastic properties, infer that: (i) the biofilm thickness increases with exposure time and declines with depth; (ii) the biofilm decreases the hydrophobic-ity and changes functional groups on a plastic surface; (iii) Alphaproteobacteria, Gammaproteobacteria, and Bacteroidia could be the core microbiotome of the PE associated biofilms; and (iv) the dominant colonizing microbial community varies during the biofilm formation.

The sticky matrix of the biofilm traps microplastic particles. The biofilm is subsequently processed and dispersed in order to release microplastic particles to

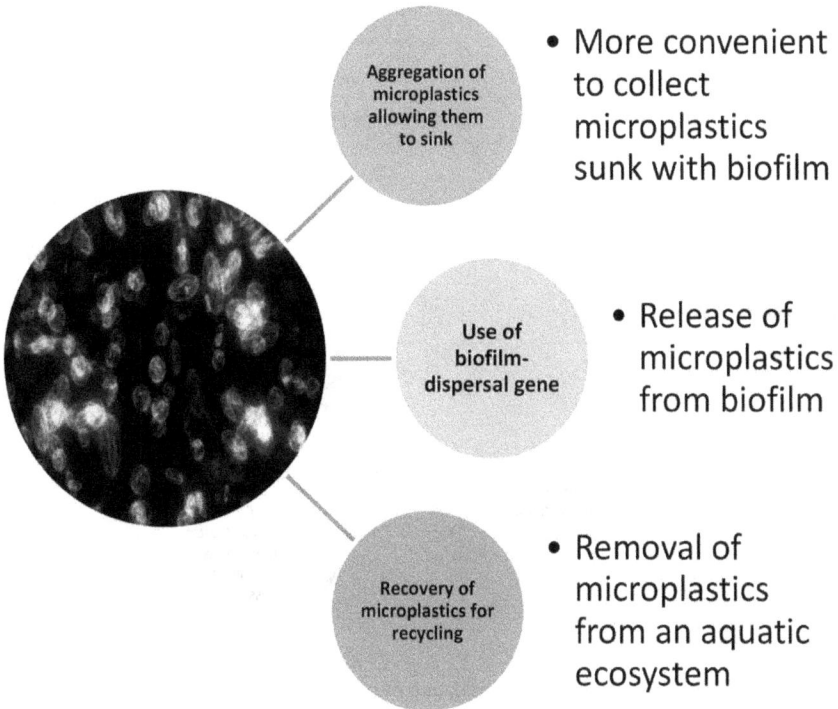

FIGURE 5.6 Application of biofilms for microplastics remediation in an aquatic environment.

follow their processing and recycling. Microbial degradation of the polymers is preferred over physical and chemical methods because it leads to complete degradation and mineralization and offers a viable and environment-friendly solution for reducing microplastics load in aquatic environments. Scientists employ biofilms in cleaning aquatic environments in three simple steps as outlined in Figure 5.6.

A biofilm offers nutrient bioavailability and sharing of metabolites without accumulation of metabolic products (Ghosh 2019). This process results in increased cell viability and degradation efficiency.

5.4   FACTORS AFFECTING THE APPLICATION

Biodegradation of polymers is affected by many factors broadly categorized by environmental conditions to which the plastics/microplastics are exposed, and the polymer characteristics (Figure 5.7). The nature of the interactions of various factors influencing the degradation processes is explained in Table 5.5.

As observed in a review study by Mohanan et al. (2020), many variables the biodegradation of a polymer is measured against include (i) substrate weight loss, (ii) alterations in a polymer's mechanical properties, (iii) chemical changes in a polymer,

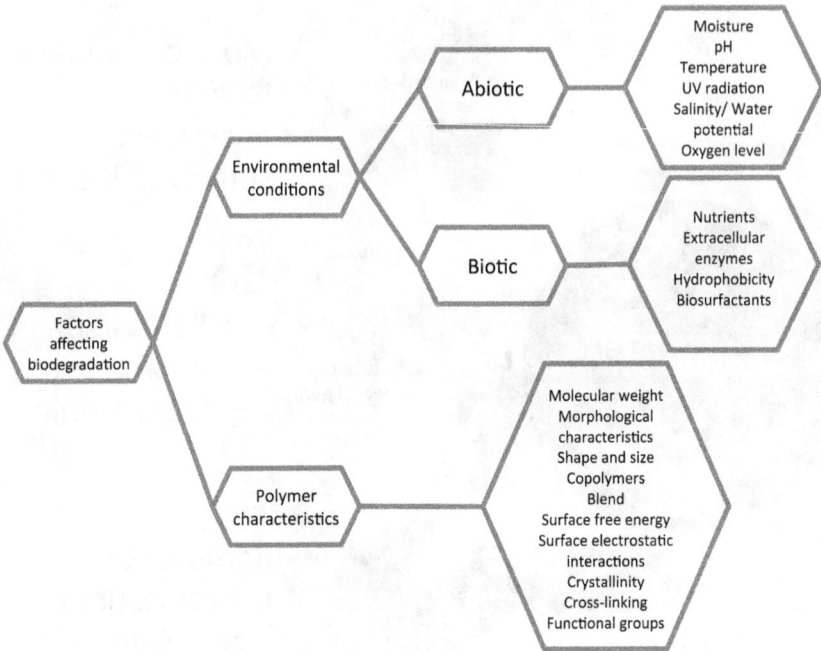

FIGURE 5.7 Factors affecting biodegradation of polymers.

and (iv) proportion of CO_2 emissions. Their study based on the reviews of earlier experiments also reveals that microbial actions on polymers bring about certain changes in the physical properties, for example tensile strength, crystallinity, and water uptake.

5.5 FUTURE PROSPECTS

Our planet in our times is facing numerous environmental and ecological problems. Microplastics in recent years have emerged as one of the most serious and formidable menaces. Although the plastics are invariably projected as recalcitrant to biodegradation, various studies, as we have observed in our discussion, suggest that they are biodegradable to various degrees. In the natural environments, especially in the hydrosphere and pedosphere, an astounding variety of microorganisms prevails. Many of the microbial species have been found to play a crucial role in various steps involved in the polymer degradation process. A bright opportunity is ahead to carry out studies on possible synergism among the microorganisms contributing to polymer degradation. Such research focus will provide an insight for making efforts towards more efficient biodegradation of various types of plastics in the future.

A few species of microorganisms tested so far for their potential of plastic degradation represent only a tip of the microbial "iceberg" of the prevailing astonishing diversity of decomposers. Researchers are challenged by the gargantuan diversity of the microorganisms thriving in diverse environments. Exploration of more and

TABLE 5.5
Various factors and their influence on polymer biodegradation.

Factors	Influence on degradation various factors and their influence on polymer biodegradation
Environmental conditions	
Moisture	• Basic necessity of the microorganisms to proliferate and carry on hydrolytic activity. • Necessary for enzyme kinetics; increases rate of hydrolysis by creating chain scission reactions.
pH	• Affects rates of hydrolytic reaction, and activity of the enzymes. • Degradation products alter pH. • Alters microbial growth rate and subsequently degradation rates.
Salinity/Water potential	• Determines water movement between the two environments: water body and microbe cells, thus affecting microbial growth and proliferation. • The higher the solute concentration beyond a limit, the lower the microbial growth in the aquatic environment, therefore lower degradation rates.
Temperature	• Softens the polymer to the extent where it undergoes degradation, breaking down into oligomers and dimers (softening temperature). • Affects degradation capacity of the microbial extracellular enzymes. • High temperature denatures enzymes, thus reducing degradability. • Polyesters with high melting points are less prone to degradation. • Polyesters with low melting points are more efficiently degraded by extracellular enzymes.
Oxygen level	• Determines microbial activity during biodegradation under aerobic conditions. • In the aerobic degradation process, O_2 is the required terminal electron acceptor. • The oxidation process enables a faster conversion of polymers into fragments. • The higher the oxygen levels in the aquatic environment, the greater the degree of degradation by aerobic decomposers.
UV radiation	• Contributes ionizing radiation playing a significant role in initiating weathering effects. • Light intensity and prolonged exposure to UV radiation accelerates plastic fragmentation and photooxidation and, thus, the degradation rates.
Nutrients	• Nutrients serve as stimulants for the microbes in the process of polymer biodegradation. • Suitable additives promote microbes' growth as well as biodegradation rates. • Some conventional polymers can also be biodegradable if they contain biodegradable additives.
Extracellular enzymes	• Extracellular enzymes produced by microbes are pivotal in the biodegradation processes. • The amount of enzymes secreted by a biofouling community targeting specific polymers and their kinetics in an appropriate environment determines the efficiency and rates of a polymer's biodegradation.
Hydrophobicity	• As a physical property of the molecules in the synthetic polymers, the hydrophobicity influences microbial actions on a plastic surface. • Cell surface hydrophobicity affects colonization and biofilm formation. • Hydrophobicity of polyethylene decreases during biofilm formation, resulting in the abundant hydrophilic C–O and C=O groups on the surface.

(Continued)

TABLE 5.5
(Continued)

Factors	Influence on degradation various factors and their influence on polymer biodegradation
Biosurfactants	• Reduce surface tension and change polymers' interfacial properties. • Enhance biodegradation processes owing to the presence of specific functional groups. • Help enhance degradation rates even under unfavorable environmental conditions (e.g., extreme temperatures and pH).
Polymer characteristics	
Molecular weight	• Plays key role in defining polymer properties. • Normally, the greater the molecular weight the lower the degradability/degradation rates. • Polymers with lower molecular weight are more prone to degradation by the enzymes produced by microbes.
Nature of plastics	• Plastics manufactured from petrochemical sources are harder to degrade on account of their hydrophobic and non-polar nature. • The hydrophobic and non-polar features of plastics resist biofilm development, thus reducing polymer degradation rates.
Morphological characteristics	• Amorphous polymers are more susceptible to microbial action and subsequent degradation than the crystalline ones. • Crystallinity-ridden polymers are very difficult to undergo degradation.
Surface free energy (SFE)	• It dictates how a solid behaves when it comes in contact with a liquid and determines the adhesion between two phases. • The SFE of a plastic type and the surface tension of the water in which the microorganisms perpetuate determine the plastic-water (with microbes) interactions. • The higher the SFE value of a plastic, the easier it will be wet by the liquid and the higher the affinity it will have for microbial attachment.
Surface electrostatic interactions (SEI)	• SEIs serve as a dominant factor in determining conformation of biomolecules and play an important role in the stability and functions of biomolecules. • Chemical reactivity, molecular recognition, and biological activity correlate with SEI potential.
Shape and size	• Larger surface area is functionally more favorable for the proliferation of microbial community. • There are greater chances of direct contact between the substrate and enzyme with smaller plastic size, thus higher rates of degradation. • Enzyme lipase acts on larger polymers with slower rates.
Crystallinity	• Slows down molecular mobility. • Crystalline phase does not adsorb water. • Significant reduction in degradation of the polymers, such as that of poly(L-lactide) (PLLA). • The higher the degree of crystallinity in a polymer, the greater the resistance against degradation.
Cross-linking	• Increases molecular weight, stiffness, waterproofness of polymers via covalent bonds between polymer chains. • Reduces mobility/viscosity, increases hydrophobicity, decreases microbial attachment, and enhances resistance to degradation.

Factors	Influence on degradation various factors and their influence on polymer biodegradation
Blend	• Mixture of two or more polymers of different physical characteristics; a blend is a new material, analogous to metal alloys. • Blending of polymers to form a tougher product results in declined degradation processes.
Copolymers	• Copolymer, a polymer made up of two or more monomer species, for example, polyethylene-vinyl acetate (PEVA), acrylonitrile butadiene styrene (ABS), and nitrile rubber, is formed through a process called copolymerization. • The degradation process is affected as per the physicochemical properties of a monomer species involved in the copolymerization.
Functional groups	• The functional groups, generally attached to the non-polar core of the carbon atoms in every repeating unit of the corresponding polymer, determine specific chemical characteristics of the polymers. • Functional polymers (i.e., the polymers bearing functional groups that have high polarity or reactivity) are likely to create more favorable conditions for polymer degradation.

more organisms to ascertain their biodegradation potential is another front for future research. A huge diversity of environments—cryospheres, marine waters, freshwater lakes, ponds, wetlands, rivers, sever lines, wastewaters, all types of soil, solid waste management sites, landfills, petroleum wastes, polymer dump sites, etc. open ample scope for the exploration of new microbial strains vital for polymer degradation. A number of new strains with very high biodegradation efficiency are likely to be registered in for future studies.

Research projects focusing on microplastics biodegradation explore the scope of incorporating non-microbial species such as small invertebrates. Some insect larvae have capability to ingest and degrade the microplastics by endoenzymes and thus can prove phenomenal in aiding in the process of polymer waste management. More scrutiny in this area needs to be done to include many more non-microbial species as potential candidates for plastic degradation.

Genetic engineering of isolated, identified, and characterized microorganisms and/ or of their enzymes for specific purpose of enhanced polymer degradation efficiency is an innovative technique likely to become popular in the future. New advances in genetic engineering, including gene editing techniques, are likely to become promising methods to reduce microplastics pollution loads in various environments.

In addition to recycling of plastic wastes through their biodegradation for use as a carbon and energy source by the microorganisms, they can be converted into useful alkane products using microbial technology. Many other alternatives of the plastic debris can be discovered. Plastic waste management, in fact, calls for carrying out multidisciplinary research projects.

The plastisphere in ocean ecosystems harbors a variety of microbes which offer research on novel isolates for molecular techniques for the characterization of polymer-degrading microorganisms and enhancing enzymatic activity levels and on the use of omics-based technology aimed at accelerating the processes leading to polymer degradation. Addition of photosensitizers or pro-oxidants species and reduction of biocides and antioxidant stabilizers may be promising towards accelerating bio-degradation processes (Sánchez et al. 2020).

Biofilm formation, an interesting phenomenon relating to polymer biodegrada-tion, holds the following future research prospects: (i) kinetics of chemical parti-tioning in connection with biofilm formation, (ii) relative importance of the multiple surface adsorption to the pollutant chemicals, organic as well as inorganic, (iii) sur-face topology vis-à-vis adsorption process, (iv) mechanisms pertaining to chemical transport by microplastics to heavy metal- and chemical-contaminated biofilms, and (v) biofilms' relations to and effects on the health of human beings and aquatic life.

The future outlook for polymer material development appears to be promising in the sense that biodegradable polymers are progressively replacing the recalcitrant ones. Research focus on the manufacture of readily biodegradable and eco-friendly polymers is a need of the hour when microplastics emanating from conventional recalcitrant plastic materials are increasingly infesting the planet's ecosystems.

One of the research imperatives emerges from the Covid-19 pandemic scenario. Manufacture of biodegradable gloves, aprons, caps, and other garments and lab coats coated with drugs that could prohibit proliferation of coronavirus is an example of the solution relating to hospital-based solid waste disposal (Visan et al. 2021). Research leading to the resolution of such public health issues gives an insight for taking up new studies in our contemporary times as it would have implications for human health in the future.

5.6 SUMMARY

Microorganisms, such as bacteria, bacterial consortia, protozoa, microalgae (diatoms, flagellates, and protists), fungi and biofilms, can play a role in the degradation of micro-plastics. Biodegradation processes involve multiple biochemical reactions and require development of functional microbial agents. (Petro-)polymers, notably PE, PS, PU, PP, PET, and PVC, are recalcitrant to natural biodegradation pathways. Microbial commu-nity composition varies in accordance to their habitat specificities. Bacteria are adapted to survive under various marine environmental conditions and can survive even under pretty inhospitable conditions, such as rock environments, glacial and sea ice, snow, lakes with permanent ice cover, permafrost soils, cloud droplets, etc.

The subsequent developments following microplastic surface attachment in the marine environment include formation of stable consortia, nutrient accumulation, protection against toxic substances and horizontal gene exchange. The microbial colonies are engaged in catalyzing the metabolic processes resulting in desorption, adsorption, and breakdown of the compounds associated with microplastics. In the process of biodegradation, the microorganisms of the biofouling community alter or transform the structure of the chemicals through the enzymes they release and metabolic activities they control. The microplastic-degrading bacteria release hydro-lytic extracellular enzymes, for example, lipase, keratinase, CMCase, xylanase, and chitinase into their surroundings which play a dramatic role in the process of micro-plastic degradation. Several bacterial strains have been discovered that demonstrate an active role in microplastic degradation. Marine microflora are reported to possess greater ability for the production of enzymes as well as protein active compounds.

Bacterial enzymes playing a key role in the processes leading to microplastic degradation include lipases, esterases, cutinases, ureases, dehydratases, proteinases, and depolymerases. Among the depolymerases are PHA depolymerases, PHB depolymerases, PLA depolymerases, and PCL depolymerases. Proteinases include proteinase K acting against PLA. A number of the strains of *Pseudomonas* and *Bacillus* carry properties of partially biodegrading a variety of petro-plastics, like PE, PP, PS, PUR, PET, and PVC.

Fungi play a degradation role through promoting formation of different chemical bonds—for example, carbonyl, carboxyl, and ester functional groups—in microplastics, thus reducing their hydrophobicity, a necessary condition contributing to enhance biochemical reactions. Fungi are predominantly responsible for degrading polyester polyurethane. Fungal species, such as *Fusarium oxysporum* and *F. soloni*, have been examined for their action upon PET. *Rhizopus delemar, R. arrhizus, Achromobacter* spp., and *Candida cylindracea* have been found producing lipases and esterases, the enzymes that aid in microplastics degradation. *Penicillium* spp., *Fusarium* spp., and *Aspergillus* spp. have been screened for their lipase production, and *P. expansum*, *P. chysogenum, A. niger*, and *Trichoderma* produce lipases in pretty high amounts. Naturally occurring fungus *Zalerion maritimum* is capable of utilizing PE.

The participating microorganisms utilize carbon and energy for their metabolism, growth, and functions. Biodegradation is an energy-releasing process. The enzymes in action on the polymer surface are external or exo-enzymes. The polymer biodegradation involves three stages: biodeterioration, biofragmentation, and assimilation. Aerobic biodegradation involves use of oxygen as an electron acceptor: $C_{polymer} + O_2 \rightarrow$ $C_{biomass} + C_{residue} + CO_2 + H_2O$. Anaerobic biodegradation involves the participation of anaerobic microorganisms in the absence of oxygen: $C_{polymer} \rightarrow C_{biomass} + C_{residue} + CH_4 +$ H_2O. Polymer biodegradation under aerobic conditions is more efficient than that under anaerobic conditions. Biodegradation of a polymer to the extent of 100% is impossible because a fraction of it is incorporated into microbial biomass $(C_{biomass})$ and residues $(C_{residues})$.

The extracellular enzymes released by the acting microbes first act upon the amorphous region and then on the crystalline region. Microplastic degradation involving a specific enzyme is a process taking place in two steps: (i) enzyme binding with the polymer substrate, and (ii) hydrolytic cleavage of the polymer. The main enzymes isolated from various fungi crucial for the degradation of a range of plastic types have been identified as nylon hydrolase, lipases, manganese peroxidase, P(3B)depolymerase, glycosidase, catalase, protease, cutinase, PCL depolymerase, etc.

Microbial communities accumulate on plastic surfaces, forming a sticky substance to trap microplastic particles, known as a biofilm. Biofilm formation facilitates the establishment of non-bacterial microorganisms, for example microalgae and microscopic fungi. A biofilm represents a unique phylogenetic, functional, and ecological entity. Hydrophobicity of polyethylene decreases during biofilm formation resulting in the abundance of hydrophilic C–O and C=O groups on the surface. The sticky matrix of the biofilm traps microplastic particles. The biofilm is subsequently processed and dispersed in order to release microplastic particles to follow their processing and recycling.

Biodegradation of polymers is affected by many factors including environmental factors (abiotic and biotic) and polymer-related features, notably molecular weight, morphological characteristics, shape and size, copolymers, blend, surface-free energy, surface electronic interactions, crystallinity, cross-linking, and functional groups. In the natural environments, especially in the hydrosphere and pedosphere, an astounding variety of microorganisms prevails. Apart from manufacturing biodegradable and environment-friendly plastics, a bright opportunity exists to carry out studies aimed at creating possible synergism among the microorganisms towards evolving rapid, more effective and remunerative polymer degradation mechanisms.

REFERENCES

Akutsu, Y., Nakajima-Kambe, T., Nomura, N. and Nakahara, T. 1998. Purification and properties of a polyester polyurethane-degrading enzyme from *Comamonas acidovorans* TB-35. *Applied and Environmental Microbiology*, 64(1):62–67. doi: 10.1128/AEM.64.1.62-67.1998.

Alshehrei, F. 2017. Biodegradation of low density polyethylene by fungi isolated from Red Sea water. *International Journal of Current Microbiology and Applied Sciences*, 6(8):1703–1709. doi: 10.20546/ijcmas.2017.608.204.

Arias-Andres, M., Klümper, U., Rojas-Jimenez, K. and Grossart H-P. 2018. Microplastic pollution increases gene exchange in aquatic ecosystems. *Environmental Pollution*, 237:253–261. doi: 10.1016/j.envpol.2018.02.058.

Artham, T. and Doble, M. 2009. Fouling and degradation of polycarbonate in seawater: Field and lab studies. *Journal of Polymers and the Environment*, 17:170–180.

Asiah, W. N., Muhamad, W., Othman, R., Shaharuddin, R. and Hasni, M. S. I. 2015. Microorganisms as plastic biodegradation agent towards sustainable development. *Advances in Environmental Biology*, 9(13):8–13. https://goo.gl/f11mzG

Austin, H. P., Allen, M. D., Donohoe, B. S., Rorrer, N. A., Kearns, F. L., Silveira, R. L., Pollard, B. C., Dominick, G., Duman, R., Omari, E. I., Mykhaylyk, V., Wagner, A., Michener, W. E., Amore, A., Skaf, M. S., Crowley, M. F., Thorne, A. W., Johnson, P. W., Woodcock, H. L., McGeeham, J. E. and Beckham, G. T. 2018. Characterization and engineering of a plastic-degrading aromatic polyesterase. *Proceedings of the National Academy of Sciences of the United States of America*, 115(19) E4350-E4357. http://doi.org/10.1073/pnas.1718804115.

Auta, H. S., Abioye, O. P., Aransiola, S. A., Bala, J. D., Chukwuemeka, V. I., Hassan, A., Aziz, A. and Fauziah, S. H. 2022. Enhanced microbial degradation of PET and PS microplastics under natural conditions in mangrove environment. *Journal of Environmental Management*, 304:114273. http://doi.org/10.1016/j.jenvman.2021.114273.

Auta, H. S., Emenike, C. U. and Fauziah, S. H. 2017. Screening of Bacillus strains isolated from mangrove ecosystems in Peninsular Malaysia for microplastic degradation. *Environmental Pollution*, 231:1552–1559.

Auta, H. S., Emenike, C. U., Jayanthi, B. and Fauziah, S. H. 2018. Growth kinetics and biodegradation of polypropylene microplastics by *Bacillus* sp. and *Rhodococcus* sp. Isolated from mangrove sediment. *Marine Pollution Bulletin*, 127:15–21.

Bajt, O. 2021. From plastics to microplastics and organisms. *FEBS Open Biology*, 11(4):954–966. doi: 10.1002/2211-5463.13120.

Barratt, S. R., Ennos, A. R., Greenhalgh, M., Robson, G. D. and Handley, P. S. 2003. Fungi are the predominant microorganisms responsible for the degradation of soil-buried polyester polyurethane over a range of soil water holding capacities. *Journal of Applied Microbiology*, 94:1–8. doi: 10.1046/j.1365-2672.2003.01961.x.

Bhatia, S. K., Gurav, R., Choi, T.-R., Jung, H.-R., Yang, S.-Y., Moon, Y.-M., Song, H.-S., Jeon, J.-M., Choi, K.-Y. and Yang, Y.-H. 2019a. Bioconversion of plant biomass hydrolysate into bioplastic (polyhydroxyalkanoates) using *Ralstonia eutropha* 5119. *Bioresource Technology*, 271:306–315. http://doi.org/10.1016/j.biotech.2018.09.122.

Bhatia, S. K., Gurav, R., Choi, T.-R., Jung, H.-R., Yang, Song, H.-S., Jeon, J.-M., Kim, J.-S., Lee, Y.-K. and Yang, Y.-H. 2019b. Poly(3-hydroxybutyrate-co-3-hydroxyhexanoate) production from engineered Ralstonia eutropha using synthetic and anaerobically digested food waste derived volatile fatty acids. *International Journal of Biological Macromolecules*, 133:1–10. http://doi.org/10.1016/j.ijbiomac.2019.04.083.

Biffinger, J. C., Barlow, D. E., Cockrell, A. L., Cusick, K. D., Hervey, W. J., Fitzgerald, L. A., Nadeau, L. J., Hung, C. S., Crookes-Goodson, W. J. and Russell, J. N. 2015. The applicability of Impranil DLN for gauging the biodegradation of polyurethanes. *Polymer Degradation and Stability*, 120:178–185. doi: 10.1016/j.polymdegradstab.2015.06.020.

Cameron, K. A., Hodson, A. J. and Osborn, A. M. 2012. Structure and diversity of bacterial, eukaryotic and archaeal communities in glacial cryoconite holes from the Arctic and Antarctic. *FEMS Microbiology Ecology*, 82(2):254–267. doi: 10.1111/j.1574-6941.2011.01277.x.

Chandra, P., Enepsa, Singh, D. P. 2020a. 22-Microplastic degradation by bacteria in aquatic ecosystem. In: Chowdhary, P, Raj, A., Verma, D. and Akhter, Y. (eds.). *Microorganisms for Sustainable Environment and Health*. doi: 10.1016/B978-0-12-819001-2.00022-X.

Chandra, P., Enespa, Singh, R. and Arora, P. K. 2020b. Microbial lipases and their industrial applications: A comprehensive review. *Microbial Cell Factories*, 19:169. doi: 10.1186/s12934-020-01428-8.

Chee, J. Y., Yoga, S. S., Lau, N. S., Ling, S. C., Abed, R. M. and Sudesh, K. 2010. Bacterially produced polyhydroxyalkanoate (PHA): converting renewable resources into bioplastics. *Current Research, Technology and Education Topics in Applied Microbiology and Microbial Biotechnology*. Mendelez Vilas. 1395–1404.

Cosgrove, L., McGeechan, P. L., Robson, G. D. and Handley, P. S. 2007. Fungal communities associated with degradation of polyester polyurethane in soil. *Applied and Environmental Microbiology*, 73:5817–5824.

Danko, A. S., Luo, M., Bagwell, C. E., Brigmon, R. L. and Freedman, D. L. 2004. Involvement of linear plasmids in aerobic biodegradation of vinyl chloride. *Applied and Environmental Microbiology*, 70(10):6092–6097. doi: 10.1128/AEM.70.10.6092-6097.2004.

Debroas, D., Mone, A. and Ter Halle, A. 2017. Plastics in North Atlantic garbage patch: A boat-microbe for hitchhikers and plastic degraders. *Science of the Total Environment*, 2017:599–600:1222–1232. Doi: 10.1016/j.scitotenv.2017.05.059.

Deguchi, T., Kitaoka, Y., Kakezawa, M. and Nishida, T. 1998. Purification and characterization of nylon-degrading enzyme. *Applied Environment Microbiology*, 64(4):1366–1371.

Du, H., Xie, Y. and Wang, J. 2021. Microplastic degradation methods and corresponding degradation mechanisms: Research status and future perspectives. *Journal of Hazardous Materials*, 418. doi: 10.1016/j.jhazmat.2021.126377.

Gajendiran, A., Khare, K., Chacko, A. M. and Abraham, J. 2016a. Fungal mediated degradation of low density polyethylene by a novel strain *Chamyleomycetes viridis* JAKA1. *Research Journal Pharmaceutical, Biological and Chemical Sciences*, 7:3123–3130.

Gajendiran, A., Krishnamoorthy, S. and Abraham, J. 2016b. Microbial degradation of low-density polyethylene (LDPE) by Aspergillus clavatus strain JASK1 isolated from landfill soil. *3 Biotech*, 6:52. doi: 10.1007/s13205-016-0394-x.

Gajendiran, A., Subramani, S. and Abraham, J. 2017. Effect of Aspergillus versicolor strain JASS1 on low density polyethylene degradation. *IOP Conf Series: Materials Science and Engineering*, 263:022038. doi: 10.1088/1757-899x/263/2/02203.

Gewert, B., Plassmann, M. M. and MacLeod, M. 2015. Pathways for degradation of plastic polymers floating in the marine environment. *Environmental Science: Processes and Impacts*, 17:1513–1521. http://doi.org/10.1039/C5EM00207.

Geyser, R., Jambeck, J. R. and Law, K. L. 2017. Production, use, and fate of all plastics ever made. *Science Advance*, 3:e1700782. doi: 10.1126/sciadv.1700782.

Ghosh, S. K., Pal, S. and Ray, S. 2013. Study of microbes having potentiality for biodegradation of plastics. *Environmental Science and Pollution Research*, 20(7):4339–4355. doi: 10.1007/s11356-013-1706-x.

Ghosh, S., Qureshi, A. and Purohit, H. J. 2019. Microbial degradation of plastics: Biofilms and degradation pathways. In: Kumar, V, Kumar, R., Singh, J. and Kumar, P. (eds) *Contaminants in Agriculture and Environment: Health Risk and Remediation*. 184–199. doi:10.26834/AESA-2019-CAE-0153-014.

Gilan, I., Hadar, Y. and Sivan, A. 2004. Colonization, biofilm formation and biodegradation of polyethylene by a strain of Rhodococcus ruber. *Applied Microbiology and Biotechnology*, 65:97–104. doi: 10.1007/s00253-004-1584-8.

Glaser, J. A. 2019. Biological degradation of polymers in the environment. In: Gomiero, A. (ed.) *Plastics in the Environment*. London: Intech Open. doi: 10.5572/intechopen.85124.

Glaser, J. A. 2020. The importance of biofilms to the fate and effects of microplastics. In: Dincer, S. (ed.) *Bacterial Biofilms*. Intech Open. doi: 10.5772/intechopen.92816.

Hadad, D., Geresh, S. and Sivan, A. 2005. Biodegradation of polyethylene by the thermophilic bacterium *Brevibacillus borstelensis*. *Journal of Applied Microbiology*, 98(5):1093–1100. doi: 10.1111/j.1365-2672.2005.02553.x.

Harrison, P., Sapp, M., Schratzberger, M. and Osborn, A. M. 2011. Interactions between microorganisms and marine microplastics: A call for research. *Marine Technology Society Journal*, 45(2):12–20. doi: 10.4031/MTSJ.45.2.2.

Howard, G. T. and Blake, C. 1998. Growth of Pseudomonas fluorescens on a polyester-polyurethane and the purification and characterization of polyurethanase-protease enzyme. *International Biodeterioration and Biodegradation*, 42:213–220. doi: 10.1016/S0964-8305(98)00051-1.

Hung, C.-S., Zingarelli, S., Nadeau, L. J., Biffinger, J., Drake, C. A., Crouch, A. L., Barlow, D. E., Russell, J. N. and Crookes-Goodson, W. J. 2016. Carbon catabolic repression and Impranil Polyurethane degradation in *Pseudomonas protegens* strain Pf-5. *Applied and Environmental Microbiology*, 82(20):6080–6090. doi: 10.1128/AEM.01448-16.

Huston, A. L., Krieger-Brockett, B. B. and Deming, J. W. 2000. Remarkably low temperature optima for extracellular enzyme activity from Arctic bacteria and sea ice. *Environmental Microbiology*, 2(4):383–388. http://doi.org/10.1046/j.1462-2920.2000.00118.x.

Iram, D., Riaz, R. and Iqbal, R. K. 2019. Usage of potential microorganisms for degradation of plastics. *Open Journal of Environment Biology*, 4(1):007–015. doi: 10.17352/ojeb.000010.

Ishii, N., Inoue, Y., Shimada, K., Tesuka, Y., Mitomo, H. and Kasuya, K. 2007. Fungal degradation of poly(ethylene succinate). *Polymer Degradation and Stability*, 92:44–52.

Jiang, X., Chen, H. Liao, Y., Ye, Z., Li, M. and Klobucar, G. 2019. Ecotoxicity and genotoxicity of polystyrene microplastics on higher plant Vicia faba. *Environmental Pollution*, 250:831–838. doi: 10.1016/j.envpol.2019.04.055.

Jumaah, O. S. 2017. Screening of plastic degrading bacteria from dumped soil area. *IOSR Journal of Environmental Science, Toxicology and Food Technology*, 11(5):93–98. doi: 10.9790/2402-1105029398.

Kettner, M. T., Rojas-Jimenez, K., Oberbeckmann, S., Labrenz, M. and Grossart, Hans-Peter. 2017. Microplastics alter composition of fungal communities in aquatic ecosystems. *Environmental Microbiology*, 19(11):4447–4459. doi: 10.1111/1462-2920.13891.

Kleeberg, I., Hetz, C., Kroppenstedt, R. M., Muller, J. and Deckwer, W.-D. 1998. Biodegradation of aliphatic-aromatic copolyesters by *Thermomonospora fusca* and other Thermophilic compost isolates. *Applied and Environmental Microbiology*, 64(5):1731–1735.

Lee, K. M., Gimore, D. F. and Huss, M. J. 2005. Fungal degradation of the bioplastic PHB (Poly-3-hydroxy-butyric acid). *Journal of Environmental Polymer Degradation*, 13(3):213–219. doi: 10.1007/s10924-005-4756-4.

Lenz, R. W. 1993. Biodegradable polymers. In Langer, R. S. and Peppas, N. A. (eds.) *Biopolymers* (Vol. 107). Berlin Heidelberg: Springer. 1–40.

Li, Z., Wei, R., Gao, M., Ren, Y., Yu, B., Nie, K. and Liu, L. 2020. Biodegradation of low density polyethylene by Microbulbifer hydrolyticus IRE-31. *Journal of Environmental Management*, 263:110402.

Lucas, N., Bienaime, C., Belloy, C., Queneudec, M. Silvestre, F. and Nava-Saucedo. 2008. Polymer biodegradation: Mechanisms and estimation techniques. *Chemosphere*, 73(4):429–442.doi: 10.1016/j.chemosphere.2008.06.064

Matavulj, M. and Molitoris, H. P. 1992. Fungal degradation of polyhydroxyalkanoates and semiquantitative assay for screening their degradation by terrestrial fungi. *FEMS Micro Lett*, 103(2–4):323–331. doi: 10.1016/0378-1097(92)90326-j.

Mohanan, N., Montazer, Z., Sharma, P. K. and Levin, D. B. 2020. Microbial and enzymatic degradation of synthetic plastics. *Front Microbiol*, 11:580709. doi: 10.3389/fmicb.2020.580709.

Mor, R. and Sivan, A. 2008. Biofilm formation and partial biodegradation of polystyrene by the actinomycete *Rhodococcuc ruber*. *Biodegradation*, 19:851–858.

Mukherjee, K., Tribedi, P., Chowdhury, A., Ray, T., Joardar, A., Giri, S. and Sil, A. K. 2011. Isolation of a *pseudomonas aeruginosa* strain from soil that can degrade polyurethane diol. *Biodegrad*, 22(2):377–388. doi: 10.1007/s10532-010-9409-1.

Müller, R.-J., Schrader, H., Profe, J., Dresler, K. and Deckwer, W.-D. 2005. Enzymatic degradation of poly(ethylene terephthalate): Rapis hydrolyse using a hydrolase from *T. fusca*. *Macromolecular Rapid Communications*, 26(17):1400–1405. doi: 10.1002/marc.200500410.

Murphy, C. A., Cameron, J. A., Huang, S. J. and Vinopal, R. T. 1996. Fusarium polycaprolactone depolymerise is cutinase. *Applied and Environmental Microbiology*, 62(2):456–60. doi: 10.1128/aem.62.2.456-460.1996.

Negoro, S., Shibata, N., Tanaka, Y., Kato, D., Takeo, M. and Higuchi, Y. 2012. Three-dimensional structure of nylon hydrolase and mechanism of nylon-6 hydrolysis. *Journal of Biological Chemistry*, 287(7):5079–5090. doi: 10.1074/jbc.M111.321992.

Nishida, T., Fujisawa, M. and Hirai, H. 2001. Degradation of polyethylene and nylon 66 by the laccase-mediator system. *Journal of Environmental Polymer Degradation*, 9(3):102–108. doi: 10.1023/A:1020472426516.

Obradors, N. and Aguilar, J. 1991. Efficient biodegradation of high-molecular-weight polyethylene glycols by pure cultures of *Pseudomonas stutzeri*. *Applied and Environmental Microbiology*, 57:2383–2388.

Paço, A., Duarte, K, da Costa, J. P., Santos, P. S. M., Pereira, R., Pereira, M. E., Freitas, A. C., Duarte, A. C. and Rocha-Santos, T. A. P. 2017. Biodegradation of polyethylene microplastics by the marine fungus Zalerion maritimum. *Science of the Total Environment*, 586:10–15. doi: 10.1016/j.scitotenv.2017.02.017

Park, S. Y. and Kim, C. G. 2019. Biodegradation of micro-polyethylene particles by bacterial colonization of a mix consortium isolated from a landfill site. *Chemosphere*, 222:527–533. doi: 10.1016/j.chemosphere.2019.01.159.

Pathak, V. M. and Navneet. 2017. Review on the current status of polymer degradation: a microbial approach. *Bioresour Bioprocess*, 4:15.10.1186/s40643-017-0145-9.

Pauli, N. C., Petermann, J. S., Lott, C. and Weber, M. 2017. Macrofouling communities and the degradation of plastic bags in the sea: An in situ experiment. *Royal Society Open Science*, 4(10):170549. doi: 10.1098/rsos.170549.

Pramila, R. and Vijaya Ramesh, K. 2011. Biodegradation of low density polyethylene (LDPE) by fungi isolated from marine water—A SEM analysis. *African Journal of Microbiology Research*, 5(28):5013–5018.

Prasath, B. B. and Poon, K. 2018. The impacts of microplastics to environment. *Journal of Environmental Hazards*, 1(1) 101.

Raddadi, N. and Fava, F. 2019. Biodegradation of oil-based plastics in the environment: Existing knowledge and needs of research and innovation. *Science of the Total Environment*, 679:148–158. doi: 10.1016/j.scitotenv.2019.04.419.

Raghul, S. S., Bhat, S. G., Chandrasekaran, M., Francis, V. and Thachil, E. T. 2014. Biodegradation of polyvinyl alcohol-low linear density polyethylene-blended plastic film by consortium of marine benthic vibrios. *Int J Environmental Science and Technology*, 11(7):1827–1834. doi: 10.1007/s13762-013-0335-8.

Roohi, Bano, K., Kuddus, M., Zaheer, M. R., Zia, Q., Khan, M. F., Ashraf, G. M., Gupta, A. and Aliev, G. 2017. *Curr Pharm Biotechnol*, 18(5):429–440. doi: 10.2174/1389201018 666170523165742.

Ruiz, C., Main, T., Hillard, N. and Howard, G. T. 1999. Purification and characterization of two polyurethanase enzymes from *Pseudomonas chlororaphis*. *International Biodeterioration and Biodegradation*, 43:43–47. doi: 10.1016/S0964-8305(98)00067-5.

Rummel, C. D., Jahnke, A., Gorokhova, E., Kuhnel, D. and Schmitt-Jansen, M. 2017. Impacts of biofilm formation on the fate and potential effects of microplastic in aquatic environment. *Environmental Science and Technology Lett*, 4(17):258–267. doi: 10.1021/acs.estlett.7b00164.

Russell, J. R., Huang, J., Anand, P., Kucera, K., Sandoval, A. G. et al. 2011. Biodegradation of polyester polyurethane by endophytic fungi. *Applied and Environmental Microbiology*, 77:6076–6084.

Sakai, K., Fukuba, M., Hasui, Y., Moriyoshi, K., Ohmoto, T., Fujita, T. and Ohe, T. 1998. Purification and characterisation of an esterase involved in degradation of poly(vinyl alcohol) by *Pseudomonas vesicularis* PD. *Bioscience, Biotechnology and Biochemistry*, 62(10):2000–2007. doi: 10.1271/bbb.62.2000.

Salta, M., Wharton, J. A., Blache, Y., Stokes, K. R. and Briand, J. F. 2013. Marine biofilms on artificial surfaces: Structure and dynamics. *Environment Microbiology*, 15(11):2879–2893.

Sánchez, C. 2020. Fungal potential for the degradation of petroleum-based polymers: An overview of macro- and microplastics biodegradation. *Biotechnology Advances*, 40:107501. http://doi.org/10.1016/j.biotechadv.2019.107501.

Sanchez, J. G., Tsuchii, A. and Tokiwa, Y. 2000. Degradation of polycaprolactone at 50°C by a thermotolerant Aspergillus sp. *Biotechnology Letters*, 22:849–853. doi: 10.1023/A:1005603112688.

Sangeetha Devi, R., Kannan, V. R., Nivas, D., Kannan, K., Chandru, S., Antony, A. R. 2015. Biodegradation of HDPE by Aspergillus spp. from marine ecosystem of Gulf of Mannar, India. *Marine Pollution Bulletin*, 96(1–2):32–40. http://doi.org/10.1016/j.marphlbul.2015.05.050.

Sekiguchi, T., Ebisui, A., Nomura, K., Watanabe, T., Enoki, M. and Kanehiro, H. 2009. Biodegradation of several fibers submerged in deep sea water and isolation of biodegradable plastic degrading bacteria from deep ocean water. *Nippon Suisan Gakkaishi*, 75(6):1011–1018.

Sekiguchi, T., Saika, A., Nomura, K. Watanabe, T., Watanabe, T., Fujimoto, Y., Enoki, M, Sato, T., Kato, C. and Kanehiro, H. 2011. Biodegradation of aliphatic polyesters soaked in deep seawaters and isolation of poly(E-caprolactone)-degrading bacteria. *Polymer Degradation and Stability*, 96(7):1397–1403. doi: 10.1016/j.polymdegradstab.2011.03.004.

Sekiguchi, T., Sato, T., Enoki, M., Kanehiro, H., Uematsu, K. and Kato, C. 2010. Isolation and characterization of biodegradable plastic degrading bacteria from deep-sea environments. *JAMSTEC-R*, 11:33–41. doi: 10.5918/jamster.11.33.

Selim, M. S., Shenashen, M. A., El-Safty, S. A., Higazy, S. A., Selim, M. M., Isago, H. and Elmarakbi, A. 2017. Recent Progress in marine foul-release polymeric nanocomposite coatings. *Progress in Materials Science*, 87:1–32. http://doi.org/10.1016.j.pmatsci.2017.02.001.

Seo, H.-S., Um, H.-J., Min, J., Rhee, S.-K., Cho, T.-J., Kim, Y.-H. and Li, J. 2007. Pseudozyma jejuensis sp. nov., a novel cutinolytic ustilaginomycetous yeast species that is able to degrade plastic waste. *FEMS Yeast Research*, 7(6); 1035–1045. http://doi.org/10.1111/j.1567-1364.2007.00251.x.

Shah, A. A., Hasan, F., Hameed, A. and Ahmed, S. 2008. Biological degradation of plastics: A comprehensive review. *Biotechnology Advance*, 26(3):246–265. doi: 10.1016/j.biotechadv.2007.12.005.

Shah, Z., Hasan, F., Krumholz, L., Atkas, D. and Shah, A. A. 2013. Degradation of polyester polyurethane by newly isolated Pseudomonas aeruginosa strain MZA-85 and analysis of degradation products by GC-MS. *International Biodeterioration and Biodegradation*, 77:114–122. doi: 10.1016/j.ibiod2012.11.009.

Shimao, M. 2001. Biodegradation of plastics. *Current Opinion Biotechnology*, 12(3):242–247. doi: 10.1016/s0958-1669(00)00206-8.

Singh, Purnima, Singh, S. M. and Dhakephalkar, P. 2014. Diversity, cold active enzymes and adaptation strategies of bacteria inhabiting glacier cryoconite holes of High Arctic. *Extremophiles*, 18:229–242. http://doi.org/10.1007/s00792-013-0609-6.

Stern, R. V. and Howard, G. T. 2000. The polyester polyurethanase gene (pueA) from *Pseudomonas chlororaphis* encodes a lipase. *FEMS Microbiology Letters*, 185:163–168. doi: 10.1111/j.1574-6968.2000.tb09056.x

Sudhakar, M. Priyadarshini, C., Doble, M., Murthy, P. S. and Venkatesan, R. 2007b. Marine bacteria mediated degradation of nylon 66 and 6. *International Biodeterioration and Biodegradation*, 60(3):144–151. doi: 10.1016/j.ibiod.2007.02.002.

Sudhakar, M., Doble, M., Murthy, P. S. and Venkatesan. 2008. Marine microbe-mediated biodegradation of low- and high-density polyethylenes. *International Biodeterioration and Biodegradation*, 61(3):203–213. http://doi.org/10.1016/j.ibiod.2007.07.011.

Sudhakar, M., Trishul, A., Doble, M., Kumar, K. S., Jahan, S. S., Inbakandan, D. Viduthalai, R. R., Umadevi, V. R., Murthy, P. S. Venkatesan, R. 2007a. Biofouling and biodegradation of polyolefins in ocean waters. *Polymer Degradation and Stability*. 92(9):1743–1752. doi: 10.1016/j.polymdegradstab.2007.03.029

Taghavi, N., Singhal, N., Zhuang, W.-Q. and Baroutian, S. 2020. Degradation of plastic waste using stimulated and naturally occurring microbial strains. *Chemosphere*, 263(11):127975. doi: 10.1016/j.chemosphere.2020.127975.

Tokiwa, Y., Calabia, B. P., Ugwu, C. U. and Aiba, S. 2009. Biodegradability of plastics. *International Journal of Molecular Sciences*, 10(9):3722–3742. doi: 10.3390/ijms10093722.

Tribedi, P. and Sil, A. K. 2014. Cell surface hydrophobicity: A key component in the degradation of polyethylene succinate by Pseudomonas sp. AKS2. *Journal of Applied Microbiology*, 116(2):295–303. doi: 10.1111/jam.12375.

Tu, C., Chen, T., Zhou, Q., Liu, Y., Wei, J., Waniek, J. J. and Luo, Y. 2020. Biofilm formation and its influences on the properties of microplastics as affected by exposure time and depth in the seawater. *Science of the Total Environment*, 734:139237. http://doi.org/10.1016/j.scitotenv.2020.139237.

Urbanek, A. K. Rymowicz, W., Strzelecki, M. C., Kociuba, W., Franczak, L. and Mironczuik, A. M. 2017. Isolation and characterization of Arctic microorganisms decomposing bioplastics. *AMB Express*, 7:148. doi: 10.1186/s13568-017-0448-4.

Urbanek, A. K., Rymowicz, W. and Mironczuk, A. M. 2018. Degradation of plastics and plastic degrading bacteria in cold marine habitats. *Applied Microbiology Biotechnology*, 102(18):7669–7678. doi: 10.1007/s00253-018-9195-y.

Vega, R. E., Main, T. and Howard, G. 1999. Cloning and expression in Escherichia coli of a polyurethane-degrading enzyme from *Pseudomonas fluorescens. International Biodeterioration and Biodegradation*, 43:49–55. doi: 10.1016/S0964-8305(98)00068-7.

Visan, A. I., Popescu-Pelin, G. and Socol, G. 2021. Degradation behavior of polymers used as coating materials for drug delivery—A basic review. *Polymer (Basel)*, 13(8):1272. doi: 10.3390/polym13081272.

Volke-Sepulveda, T., Saucedo-Castaneda, G., Gutierrez-Rojas, M., Manzur, A. and Favela-Torres, E. 2001. Thermally treated low density polyethylene biodegradation by Penicillium pinophilum and Aspergillus niger. *Journal of Applied Polymer Science*, 83(2):305–314. doi: 10.1002/app.2245.

Wang, L, Tong, X., Zhu, J., Zhang, W., Niu, L. and Zhang, H. 2021. Bacterial and fungal assemblages and functions associated with biofilms differ between diverse types of plastic debris in freshwater system. *Environmental Research*, 196:110371. doi: 10.1016/j. envres.2020.110371.

Wang, X., Bolan, N., Tsang, D. C. W., Sarkar, B., Bradney, L. and Li, Y. 2021. A review of microplastics aggregation in aquatic environment: Influence factors, analytical methods, and environmental implications. *Journal of Hazardous Materials*, 402:123496. doi: 10.1016/j.jhazmat.2020.123496.

Yadav, A. N., Verma, P., Kumar, V., Sachan, S. G. and Saxena, A. K. 2017. Extreme cold environments: A suitable niche for selection of novel psychrotrophic microbes for biotechnological applications. *Applied Microbiology and Biotechnology*, 2(2):555584. AIBM.10.19080/AIBM.2017.02.555584.

Yamada-Onodera, K., Mukumoto, H., Katsuyaya, Y., Saiganji, A. and Tani, Y. 2001. Degradation of polyethylene by a fungus, Penicilline simplicissimum YK. *Polymer Degradation and Stability*, 72(2):323–327. doi: 10.1046/j.1365-2672.2003.01961.x.

Yang, J. Yang, Y., Wu, W. M., Zhao, J. and Jiang, L. 2014. Evidence of polyethylene biodegradation by bacterial strains from the guts of plastic-eating waxworms. *Environmental Science and Technology*, 48:13776–13784. http://doi.org/10.1021/es504038a.

Yang, S. S., Ding, M. Q., He, L. Zhang, C. H., Li, Q. X., Xing, D. F. and Wu, W. M. 2021. Biodegradation of polypropylene by yellow mealworms (*Tenebrio molotor*) and superworms (*Zophobas atratus*) via gut-microbe-dependent depolymerisation. *Science of the Total Environment*, 756:144087.

Yoon, M. G., Jeon, H. J. and Kim, M. N. 2012. Biodegradation of polyethylene by a soil bacterium and AlkB recombinant cell. *Journal of Bioremediation and Biodegradation*, 3:145. doi: 10.4172/2155-6199.1000145.

Yoshida, S., Hiraga, K., Takehana, T., Taniguchi, I. Yamaji, H., Maeda, Y., Toyohara, K., Miyamoto, K., Kimura, Y. and Oda, K. 2016. A bacterium that degrades and assimilates poly(ethylene terephthalate). *Science*, 351(6278):1196–1199. doi:10.1126/science. aad6359.

Yu, Y., Zeng, Y. and Chen, B. 2009. Extracellular enzymes of cold-adapted bacteria from Arctic sea ice, Canada Basin. *Polar Biology*, 32:1539–1547. http://doi.org/10.1007/ s00300-009-0654-x.

Yuan, J., Ma, J., Sun, Y., Zhou, T., Zhao, Y. and Yu, F. 2020. Microbial degradation and other environmental aspects of microplastics/plastics. *Science of the Total Environment*, 715. doi: 10.1016/j.scitotenv.2020.136968

Zhang, J., Gao, D., Li, Q., Zhao, Y., Lin, H., Bi, Q. and Zhao, Y. 2020. Biodegradation of polyethylene microplastic particles by the fungus *Aspergillus flavus* from the guts of wax moth *Galleria mellonella*. *Science of the Total Environment*, 704:135931. doi: 10.1016/j. scitotenv.2019.135931.

6 Potential Removal of the Microplastics in Marine Environment by Membrane Technology
Limitations and Future Solutions

Reverse osmosis (RO), ultrafiltration (UF), dynamic membranes (DM), and membrane bioreactors (MBR) represent the major membrane technologies somewhat distinctive in their role and performance towards the removal of microplastics from water/wastewater. Widely applied worldwide, the MBR technology has emerged as the most powerful and efficient membrane technology. Polymeric membranes have the ability to separate suspended particles, salts, microplastics, etc. and even the smallest particles (protozoa, bacteria, and viruses). Wastewater and industrial effluents treatment using appropriate membrane technologies would help reduce microplastics in the marine environment. Reuse and recycling of the materials generated from polymeric membranes is necessary. Recycled membranes can be applied in wastewater tertiary treatment, pretreatment of the RO process, and in softening of brackish water. Designed technologies should be efficient in selectively removing different sizes of the plastics. Technological innovations should be coupled with the implementation of global legislations and production of bio-based polymer items in the future.

Microplastics in the marine environment can be significantly reduced by using available membrane technology. The wastewaters and industrial effluents being drained into waterways without treatment and after improper treatment have heavy loads of microplastics which eventually find the oceans as their ultimate home where they complete their subsequent cycles in the process of which they impact the physical environment, marine biota, and human health and also get linked with terrestrial life. Membrane technology could help to break down this vicious cycle to a great extent. Drinking water supply systems in human settlements can also supply microplastic-free water using membrane technology. The same can also be used at home to avail clean water.

DOI: 10.1201/9781003312086-6

6.1 USE OF MEMBRANE TECHNOLOGY TO ADDRESS MICROPLASTIC POLLUTION

Microplastic flux into marine waters cannot be effectively controlled even if the wastewater and effluent treatment involving physical and chemical methods are adopted. Membrane technology offers a concrete solution to control microplastic pollution. In the integrated wastewater treatment, membrane technology is applied as a tertiary effluent treatment, especially for retaining small-sized plastic particles of less than 100 nm. Although primary and secondary treatment processes in the conventional wastewater treatment methods also efficiently contribute to the removal of microplastics (Talvitie et al. 2017), a smaller proportion of the microplastics in the effluent goes on adding microplastics in the aquatic environment it drains into. Advanced technologies, including membrane technology, however, offer appreciable solutions of complete microplastic removal from the final effluent to maintain the quality of the aquatic environment to which the treated wastewater makes its way.

Microplastic pollution culminating into environmental disruptions, extreme ecological risks, and human health hazards is the major issue facing our contemporary world. It is also not probable—with such intensive and extensive use of conventional plastic products in the whole world and with the plastic manufacturing industry rampantly engaged in the production and marketing of these plastic products—that microplastic emissions can be halted to a significant level despite the best management in practice. Under such a prevailing gloomy environmental scenario, membrane technology can emerge as a panacea. This technology, in fact, is already working with promises to address plastic pollution, but at a slow pace.

6.1.1 REVERSE OSMOSIS

Reverse osmosis (RO) is based on the principle of the reversal of osmosis, which is the passing on of a solvent (pure or pollutant-free water) across a membrane in the direction opposite to natural osmosis, by applying hydrostatic pressure greater than that of osmotic pressure. The RO technology is generally applied for pulling pure water out of polluted or salted water and in this process water is also freed from other pollutants, like heavy metals, and other contaminants including microplastics. RO technology is generally used in the water treatment systems operating in municipalities and industrial units using nanofiltration membranes with pore size more than 2 nm. A high pressure (10–100 bar) to a concentrated water solution (containing impurities including microplastics) is applied that helps force the water to pass through a semipermeable membrane, leaving the impurities in a more concentrated water solution (Poerio et al. 2019).

RO technology for recovering pure usable water out of the polluted water that is unusable for domestic purposes and in industrial production processes is also being commonly used in food and beverage production, pharmaceutical manufacturing and is also being applied for desalination of seawater (Antony et al. 2011; Poerio et al. 2019).

In applying RO technology, the major problem encountered is that of membrane fouling that leads to reduced RO efficiency as well as RO equipment durability. To overcome this problem to a certain extent, a pretreatment stage is required which generally involves the usage of certain chemicals like coagulants, oxidizing agents, disinfectants, and antiscalants (Jiang et al. 2017; Goh et al. 2018). Some other membrane fouling mitigation techniques include surface modification, cleaning, and use of novel membrane materials. Among them the novel membrane materials have high potential for controlling membrane fouling effectively. However, complete prevention of fouling is not attainable (Jiang et al. 2017).

Advances in ultrafiltration (UF) membrane technologies for application for wastewater treatment have led to their use in membrane pretreatment in seawater desalination plants. This technique results in steady performance of the desalination process as revealed through water quality and influx (Lau et al. 2014). The RO-UF desalination plants at industrial scale are operating in several countries with satisfactory performance. There are 15,906 operational desalination plants across the world's 177 countries producing approximately 95 million m³ of water per day for human use (Jones et al. 2019).

6.1.2 ULTRAFILTRATION

Ultrafiltration (UF) is one of the most simple, feasible, and cost-effective water treatment methods used, especially for recovering clean drinkable water. The other qualifications of the method are low energy expenditure, high separation efficiency, and compact plant size (Moslehyani et al. 2019; Poerio et al. 2019). This technique requires low pressure (1–10 bar) to be applied for the UF membrane (pore size between 1–100 nm) to discard particulates; suspended solids; harmful organisms like protozoa, bacteria, and viruses; and biomolecules such as proteins and fats. The other achievable water quality parameters following the UF technique are declined BOD (biochemical oxygen demand) to the extent of 95% or more, removal of harmful pathogens up to 90–100%, and significant reduction in turbidity (Poerio et al. 2019).

Some of the parasitic organisms present in contaminated water, *Giardia* (responsible for the diarrheal disease giardiasis) and *Cryptosporidium* (responsible for cryptosporidiosis, a respiratory and gastrointestinal illness) among them, may cause serious health problems to the people dependent on such water. UF treatment can make water free from such disease-causing organisms. UF, thus, can tackle the secondary and tertiary filtration techniques used in conventional wastewater treatment avoiding sedimentation, flocculation, and coagulation (secondary stage) and sand filtration and chlorination (tertiary stage) and such replacement is followed by some municipalities employing the UF treatment (Figure 6.1).

Several industries during the production processes they employ consume enormous amounts of water and discharge huge amounts of polluted and toxic wastewater in natural water bodies. Industries such as those manufacturing chemicals, paper, textiles, steel, etc. can reuse the wastewater for their specific purposes by employing the UF treatment technique.

Efficiency of the UF treatment method for the removal of lower molecular weight organic matter is low. The organic matter removal, however, is ensured by coupling the UF method with the coagulation step and this technology is widely applied nowadays. This technology is less efficient for making the final effluents free from

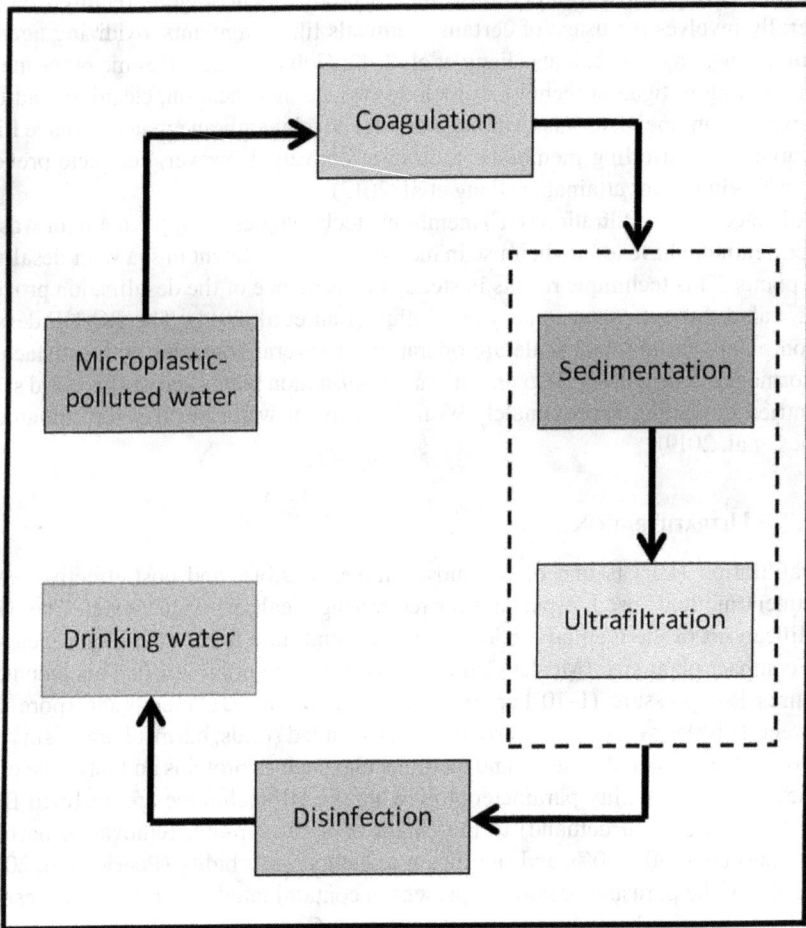

FIGURE 6.1   Schematic representation of ultrafiltration for microplastic removal.

Source: Adapted from Poerio et al. (2019).

microplastic particles (Mason et al. 2016; Talvitie et al. 2017). Microplastics originating from polyethylene (PE) make up the largest proportion in wastewater. PE
density ranging between 0.92 and 0.97 g cm^{-3} is quite close to that of water, due
to which their removal during the treatment process is pretty difficult. PE removal
efficiency after coagulation is less than 15%, which, in case of small-particle-size PE
(d < 0.5 mm), has been found significantly increased from 13 to 91%, when polyacrylamide (PAM) was added for enhancing the coagulation process (Poerio et al. 2019).

General principles relating to the UF plastic characteristics, like size, shape. and
chemical composition, affect the UF performance. Larger PE particles might induce
membrane fouling and, thus, negatively influence PE performance. Larger proportion
of PE in fiber shape might not be retained in the water treatment process. Substantial
improvements in the design of the final stage of treatment would be necessary for

enhancing PE removal efficiency in the UF process, which, as a general principle, can be instrumental in complete removal of PE particles.

6.1.3 Dynamic Membrane Technology

Dynamic membrane technology bases itself on the formation of a cake layer called dynamic membrane (DM). The DM functions as a secondary membrane which creates a barrier for the particles, such as those of the plastics, and other foulants in the wastewater, preventing them from filtering beyond the supporting membrane. This technology is quite different from the UF technology in the sense that in the DM process a resistance against the filtration of plastic particles is created by means of the cake layer. Formation of a thicker layer and dense fouling may cause hindrance in the mechanism and diminish the membrane performance.

There are some specific parameters the formation process relies upon. These are related to membrane material and pore sizes of the supporting membranes, particle size and concentrations of the deposited materials, and the conditions related with process operations, (e.g., pressure and cross-flow velocity) (Ma et al. 2013).

DM technology offers many advantages over the others. Some examples of special attributes pertaining to DM technology are compact treatment set-up, use of low-cost material, no use of extra chemicals, and low-energy inputs needed for operations. As the DM technology is applicable for the low-density particles that do not easily settle down, it is appropriate for the removal of microplastics in the wastewater. This technology in recent years has attracted municipalities and industries for wastewater treatments, especially for surface water, oily water, industrial wastewater and sludge.

6.1.4 Membrane Bioreactor

A membrane bioreactor (MBR) is a system involving catalysis enhanced by biological catalysts (viz., bacteria and enzymes) and coupled to a separation process operated by a membrane system, in general microfiltration or ultrafiltration (Xiao et al. 2019).

Membranes are easily used to make up different compartments due to which a system involving heterogeneous reactions can be devised. The heterogeneous reaction system encompasses (organic/water)/multiphase (liquid/gas) reactions. The different phases, interestingly, may be kept separate (e.g., in case of a membrane-based solvent extraction process), or they may be dispersed into each other (e.g., in case of a membrane emulsification process). This technology can well be integrated with other processes (e.g., RO) in consonance with green chemistry, spelling out many promising attributes, including product quality improvement and product novelty, in addition to positive impacts on the environment (Judd 2016).

MBR technology in recent years has emerged as one of the most powerful technologies adopted worldwide for municipal and industrial wastewater treatment with recorded high efficiencies. This technology is also being extended to many new fields, like foods, pharmaceuticals, biorefineries, and biodiesel production.

Applications of MBR technology in microplastic treatment requires decrease of solution complexity by organic matter biodegradation. This leads to the purification of microplastics and their further treatment. The membrane process allows the

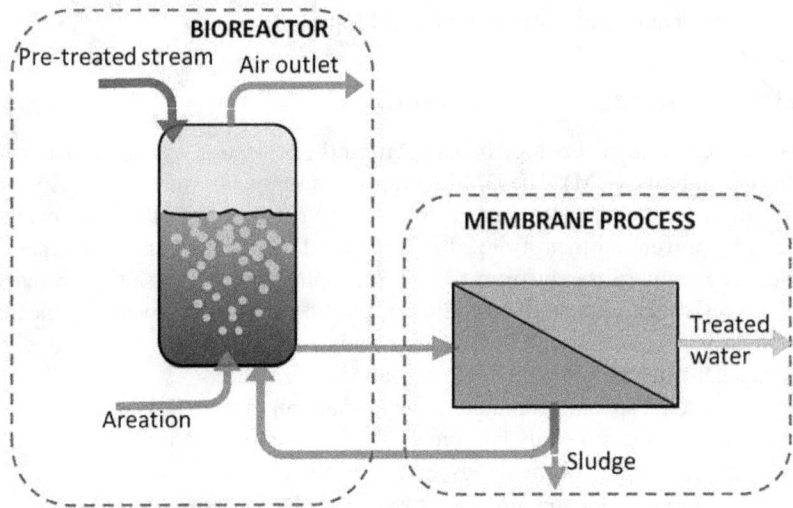

FIGURE 6.2   Schematic representation of an MBR process.

Source: Poerio et al. (2019).

microplastics to become concentrated in the retentate stream via a few steps, namely entering of the pre-treated stream into the bioreactor for organic matter biodegradation, and pumping of the produced mixed liquor along with a semi-cross flow filtration system for the separation process (Figure 6.2). Compared to other processes, the MBR demonstrated the most efficient performance by removing microplastics up to 99% (Poerio et al. 2019).

A wide variety of bacteria and mycetes, as Poerio et al. (2019) documents, can degrade phthalate esters completely. The biodegradation efficiency of MBR depends on the physicochemical properties of the phthalate esters, as well as on operational conditions, like hydraulic retention time (HRT) and initial feed concentration, etc. A novel bacterium, *Idonella sakaiensis*, has the ability to utilize polyethylene terephthalate (PET) as a major carbon and energy source (Yoshida et al. 2016). The bacterium secretes two enzymes that may efficiently convert PET into less harmful monomers: terephthalic acid and ethylene glycol. Associated with MBR, this novel bacterium can have a crucial role in the removal of harmful plastic particles.

6.1.5   POLYMERIC MEMBRANES

Polymeric membranes, or organic membranes, constitute a family of membranes produced from polymers. Such a membrane is a semi-permeable filter media applied for pressure-driven water/wastewater treatment. Their pore sizes can be easily controlled during their formation. The most important feature required in the polymeric membranes is their affinity to a particular component. A very small space along with high flexibility is needed for their installation. A polymer is chosen in accordance with the requirement as per the objective of the separation task. The polymeric membranes can be conveniently operationalized in microfiltration, ultrafiltration, nanofiltration, and reverse osmosis and carry the ability to separate suspended particles, salts, microplastics, etc. and even the smallest particles, including protozoa, bacteria, and viruses.

Some of the widely used polymeric membranes include PE, polycarbonate, polypropylene, polyimide, polysulfone, polyethersulfone, polyacrylonitrile, cellulose acetate, and polytetrafluoroethylene. Some of the polymeric membranes, such as polyvinylidene fluoride ultrafiltration membranes, have embedded nanomaterials, such as metal/metal oxide or carbon nanotubes to enhance the polymeric membrane performance. Cost-effectiveness and wide range applicability of the polymeric membranes prompts the separation industry to make them their priority for many tasks. Further, these membranes are quite popular in separation science and technology in the academics.

6.2   HOW MEMBRANE TECHNOLOGY CAN REDUCE MICROPLASTICS IN MARINE ENVIRONMENT

As mega-, macro-, and mesoplastics get fragmented into microplastics in the marine environment, the first and the foremost approaches to reduce microplastics is to prevent the plastic wastes from entering oceans. Proper disposal of the plastics wastes in the localities themselves would be the first step in this direction. A cleanup drive of rivers to make them free from plastics and thus not allowing the plastic wastes to end up in oceans is another key preventive approach. The fragmentation processes of larger plastic particles due to a combination of strong UV radiation and physical abrasion by waves—and, thus creation of microplastics—is most effective on beaches. Therefore, cleaning up of the beaches can make a significant difference in reducing microplastics load in the marine environment.

An enormous load of plastic wastes, much of which fragmented into microplastics in the human settlements on land, is received in marine ecosystems through wastewaters directly from the coastal areas or via rivers and streams from distant inhabited areas. If all the wastewater and industrial effluents are treated using effective membrane technologies and microplastics are separated at the source, the marine environment would not be fed by these dangerous pollutants at the rates being currently witnessed.

Marine waters are freed from microplastics in seawater treatment and drinking water supply systems involving desalination and other membrane-based treatment systems. The microplastics separated in the process are squeezed out of the marine waters and can further be disposed of following a safe, environment-friendly approach.

Preventive measures are easier, more cost-effective, and practically more relevant from every sense compared to the measures for *in situ* microplastic reduction in the marine environment. Handling the vastest, most complex, and largely unmanageable marine ecosystems for microplastic pollution mitigation and/or microplastic reduction would be pretty difficult, even unthinkable. There is an appreciable research approach based on the concept of Herbort and Schuhen for freshwater systems and simultaneous development of add-on technology for ocean water utilization processes. In this, silane-based microplastic agglomerates are formed as per the cloud point principle through the application of special organosaline-based precursors, which, via Van der Waals forces, have a high affinity to unreactive microplastics (Schuhen et al. 2018).

6.3 REUSE AND RECYCLING OF POLYMERIC MEMBRANES OF BIOREACTORS

Polymer recycling enabling polymer reuse is one of the environmentally sound ways to diminish environmental crises caused by continuous polymeric waste accumulation generated from overwhelmingly used polymer products in day-to-day life. In our contemporary world haplessly revealing a dismal scenario of water pollution, worldwide applications of membrane technologies for water and wastewater treatment are being increasingly witnessed. Such a vast use of the membranes in the filtration processes is also leaving behind huge amounts of polymer wastes that also draw our attention for their proper management. Development of effective technical approaches is needed to phenomenally manage, reuse, and recycle the materials generated from polymeric membranes.

The membrane-based technologies have also captured a well-established global market. It is very likely that in the future we shall witness more extensive use of the membrane technologies with more innovative modules in a health-conscious world, with a corresponding increase in waste materials calling for effective management focusing on reusing and recycling of polymeric membranes. The membrane recycling processes contribute to reducing greenhouse gases as well as the consumption of non-renewable resources, which is helpful in reducing the global impact of desalination on the environment.

6.4 EXAMPLES

Many examples relating to an increasing trend in the recycling and reuse of polymeric membranes, as also documented by Poerio et al. (2019), have come to the fore. One of the notable examples emanates from LIFE+ TRANSFOMEM Project with a focus on the transformation of disposed RO membranes into recycled ultra- and nanofiltration membranes. In this project, experiments were conducted at laboratory and pilot levels suggesting that the membrane technology could well be upgraded to a circular economy system. Exposure of the membranes to a concentrated solution of free chlorine was the basis of transforming the membranes into ultrafiltration membranes following the complete removal of the polyamide layer. Two different membrane transformation approaches—active and passive—have come into effect. Recycled membranes developed in the LIFE+ TRANSFOMEM projects can be effectively applied in (i) wastewater tertiary treatment, (ii) pretreatment of the RO process, and (iii) softening of brackish water. The recycling and reuse processes devised in the project can assure recycling of about 70% of the membranes and the befitting reuse of these membranes can save 85 to 95% in comparison to the acquisition of new commercial membranes (Transfomem Project 2018).

Recycling and reuse of water treatment RO membranes upon the expiry of their life is also carried out by a German firm MemRe (www.memre.de/). Such facilities can take care of huge loads of the used-up membranes (estimated at around 100,000 tons per year with natriumchlorid, sludges, and, in a few cases, also radio nuclides) in the environment, about 65% of which is used to refurbish operating plants, according to the information provided by MemRe. This amount of membrane wastes is apart from that generated from ultra- and nanofiltration membranes.

Interestingly, the focus of polymer production is now increasingly shifting to the use of biodegradable and recyclable materials derivable from bio-based polymers which are sustainable alternatives to petrochemical-based conventional plastic products.

6.5 LIMITATIONS OF CURRENT METHODS AND INFRASTRUCTURE

Water/wastewater treatment technologies as of today are quite effective in removing almost all pollutants and thus turning water into its pure drinkable form. However, some drawbacks are encountered when micro- and nanoplastics separation is taken into account. Wastewater treatment plants (WWTPs) and concerned industrial units are not properly and adequately equipped with the technologies effectively dealing with the microplastics in effluents. Looking at the all-pervading microplastics in the water resources of the planet, development of designed microplastic treatment processes is undoubtedly a pressing need to combat the problem and provide a sound, sustainable supply of healthy, microplastic-free water on a continuing basis.

Designed technologies should be efficient in selectively removing different sizes of plastics (e.g., meso-, micro-, and nanoparticles). It has been found that the advanced tertiary stage processes in the wastewater treatment processes require removal of microplastics. It is also observed that among the tertiary processes, the MBR-associated membrane processes are most efficient in the handling of microplastics in wastewater, contributing to the removal of almost 100% microplastics, and that this process also has the possibility to decrease the number of process stages in the WWTPs (Poerio et al. 2019). Therefore, MBR needs to be more extensively used in the WWTPs and in industries.

Physicochemical characteristics of the plastics would be instrumental for choosing suitable techniques ensuring a more efficient removal of plastics from the industrial effluents. Since the nanoplastics are of serious ecological, environmental, and health-related consequence, separation of these smallest plastic particles should be mandatory in the more advanced technologies. Finally, sound and safe ecological approaches need to be developed for effective cleaning of microplastics directly from the open oceans at meso and macro scales.

6.6 FUTURE SOLUTIONS AND SCOPE

Microplastics in the marine environment cannot be allowed to stay and keep expanding their volume, for their hazardous impacts on the life and the living planet would be too profound to be easily tackled in the future. Global future strategies to

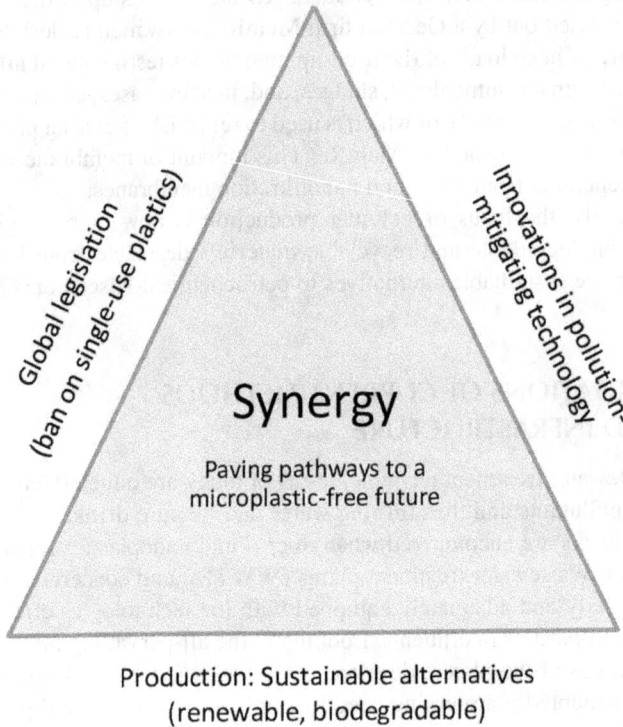

FIGURE 6.3   The pyramid of synergy. Bio-based sustainable alternatives to the conventional plastics making the base of the pyramid, and innovative mitigation technologies and a workable global legislation making the two other arms of the pyramid would create synergies paving pathways to a microplastic-free future.

control and regulate plastics must inevitably include the technologies, essentially the membrane-based technologies, for effectively limiting the microplastic pollution at the source, away from the marine ecosystems. Future WWTPs are most likely to have innovatively upgraded designs capable of processing all types and all sizes of plastic particles with 100% removal efficiency. All municipalities and other small, medium, and large human settlement areas would have to implement and ensure water/wastewater treatment with no loose point for the microplastics and nanoplastics to elute in the treated water. Desalination systems would also be needed to be more efficient and enduring. Technological innovations would have to be coupled with local, regional, and global legislations/policies and with a vision of eco-friendly alternatives in the future.

The three-dimensional strategic framework, represented as a synergy pyramid (Figure 6.3), would generate the most required synergy for striking a balance between environmental quality and socioeconomic development processes, and create pathways towards a future free from microplastic pollution.

6.7 SUMMARY

Membrane technologies, especially those represented by reverse osmosis (RO), ultrafiltration (UF), dynamic membranes (DM), and membrane bioreactor (MBR) technology, offer a concrete solution to controlling microplastic pollution. RO technology is generally used in the water treatment systems operating in municipalities and industrial units using nanofiltration membranes. The major problem of fouling encountered can be overcome through a pretreatment stage using coagulants, oxidizing agents, disinfectants, and antiscalants. UF, a simple, feasible, and cost-effective water treatment method, can discard particulates; suspended solids; harmful organisms like protozoa, bacteria, and viruses; and biomolecules, such as proteins and fats, and can tackle the secondary and tertiary filtration techniques used in conventional wastewater treatment. DM technology bases itself on the formation of a cake layer that functions as a secondary membrane creating a barrier for the particles, such as those of the plastics, and other foulants in the wastewater, preventing them from filtering beyond the supporting membrane. It is applicable for the low-density particles and, thus, is appropriate for the removal of microplastics in the wastewater. MBR is a system involving catalysis enhanced by biological catalysts, namely bacteria and enzymes, and coupled to a separation process operated by a membrane system, in general microfiltration or ultrafiltration. Having emerged as one of the most powerful technologies adopted worldwide for municipal and industrial wastewater treatment with recorded high efficiencies, the MBR technology is now also being extended to many new fields, like foods, pharmaceuticals, biorefineries, and biodiesel production. Polymeric membranes constitute a family of membranes produced from polymers. The most important feature required in the polymeric membranes is their affinity to a particular component. They carry the ability to separate suspended particles, salts, microplastics, etc. and even the smallest particles, including protozoa, bacteria, and viruses.

Prevention of the plastic wastes from entering oceans should be the first priority to reduce microplastics in the marine environment. A cleanup drive of rivers and beaches to make them free from plastics and thus not allowing the plastic wastes end up into oceans is another key preventive approach. If all the wastewater and industrial effluents are treated using effective membrane technologies, and microplastics are separated at source, marine environment would not be fed by these dangerous pollutants.

Effective technical approaches are needed to phenomenally manage, reuse, and recycle the materials generated from polymeric membranes. A notable example emanates from LIFE+ TRANSFOMEM Project with a focus on the transformation of disposed RO membranes into recycled ultra- and nanofiltration membranes. Recycled membranes can be effectively applied in wastewater tertiary treatment, pretreatment of the RO process, and for softening of brackish water. The recycling and reuse processes can assure recycling of about 70% membranes and these membranes can save 85 to 95% in comparison to the acquisition of new commercial membranes.

Designed technologies should be efficient in selectively removing different sizes of the plastics. Advanced tertiary stage processes in the wastewater treatment processes require removal of microplastics. Among the tertiary processes, the MBR-associated membrane processes are most efficient in the handling of microplastics in wastewater, contributing to the removal of nearly 100% microplastics. Physicochemical characteristics of the plastics would be instrumental for choosing suitable techniques ensuring a more efficient removal of plastics from the industrial effluents. Global future strategies to control and regulate plastics must inevitably include the technologies for effectively limiting the microplastic pollution at the source, away from the marine ecosystems. Future WWTPs are most likely to have innovatively upgraded designs capable of processing all types and all sizes of plastic particles with 100% removal efficiency. Technological innovations need to be coupled with global legislations and with a vision of eco-friendly alternatives in the future. The three-dimensional strategic framework, represented as a synergy pyramid, would help create pathways towards a future free from microplastic pollution.

REFERENCES

Antony, A., Low, J. H., Gray, S., Childress, A. E., Le-Clech, P. and Leslie, G. 2011. Scale formation and control in high pressure membrane water treatment systems: A review. *Journal of Membrane Science*, 383(1–2):1–16. doi: 10.1016/j.memsci.2011.08.054.

Goh, P. S., Lau, W. J., Othman, M. H. D. and Ismail, A. F. 2018. Membrane fouling in desalination and its mitigation strategies. *Desalination*, 425:130–155. doi: 10.1016/j.desal.2017.10.018.

Jiang, S., Li, Y. and Ladewig, B. P. 2017. A review of reverse osmosis membrane fouling and control strategies. *Science of the Total Environment*, 595:567–583. doi: 10.1016/j.scitotenv.2017.03.235.

Jones, E., Qadir, M., van Vliet, M. T. H., Smakhtim, V. and Kang, S.-m. 2019. The state of desalination and brine production: A global outlook. *Science of the Total Environment*, 657:1343–1356. doi: 10.1016/j.scitotenv.2018.12.076.

Judd, S. J. 2016. The status of industrial and municipal effluent treatment with membrane bioreactor technology. *Chemical Engineering Journal*. 305:37–45. doi: 10.1016/j.cej.2015.08.141.

Lau, W. J., Goh, P. S., Ismail, A. F. and Lai, S. O. 2014. Ultrafiltration as a pretreatment for seawater desalination: A review. *Membrane and Water Treatment*, 5(1):15–29. Doi: 10.12989/mwt.2014.5.1.015.

Ma, J., Wang, Z., Xu, Y., Wang, Q., Wu, Z. and Grasmick, A. 2013. Organic matter recovery from municipal wastewater by using dynamic membrane separation process. *Chemical Engineering Journal*. 219:190–199. doi: 10.1016/j.cej.2012.12.085.

Mason, S. A., Garneau, D., Sutton, R., Chu, Y., Ehmann, K., Barnes, J., Fink, P., Papazissimos, D. and Rogers, D. L. 2016. Microplastic pollution is widely detected in US municipal wastewater treatment plant effluent. *Environmental Pollution*, 218:1045–1054. doi: 10.1016/j.envpol.2016.08.056.

MemRe. n.d. RO membrane recycling: We are your global partner for your membrane recycling. *MemRe*. www.memre.de/ (accessed on May 13, 2022).

Moslehyani, A., Ismail, A. F., Matsuura, T., Rahman, M. A. and Goh, P. S. 2019. Recent progress of ultrafiltration (UF) membranes and processes in water treatment. In: Ismail, A. F., Rahman, M. A., Othman, M. H. D. and Mastuura, T. (eds.) *Membrane Separation Principles and Applications*. Amsterdam: Elsevier Inc. 85–109.

Poerio, T., Piacentini, E. and Mazzei, R. 2019. Membrane processes for microplastic removal. *Molecules*, 24(22):4148. doi: 10.3390/molecules24224148.

Schuhen, K., Sturm, M. T. and Herbort, A. F. 2018. Plastics in the environment. In: Gomiero, A. (ed.) *Technological Approaches for the Reduction of Microplastic Pollution in Seawater Desalination Plants and for Sea Salt Extraction*. London: Intech Open. doi: 10.5772/intechopen.81180.

Talvitie, J., Mikola, A., Koistinen, A. and Setälä, O. 2017. Solutions to microplastic pollution—Removal of microplastics from wastewater effluent with advanced wastewater treatment technologies. *Water Research*, 123:401–407. doi: 10.1016/j.watres.2017.07.005.

Transfomem Project. 2018. Achievements and conclusions of the Life Transfomem Project on the recycling of disposed membranes. www.water.imdea.org/news/2018/achieve-ments-and-conclusions-life-transfomem-project-recycling-disposed-membranes (accessed on May 13, 2022).

Xiao, K., Lianga, S., Wanga, X., Chena, C. and Huanga, X. 2019. Current state and challenges of full-scale membrane bioreactor application: A critical review. *Bioresource Technology*, 271:473–481. doi: 10.1016/j.biortech.2018.09.061,

Yoshida, S., Hiraga, K., Takehana, T., Taniguchi, I. Yamaji, H., Maeda, Y., Toyohara, K., Miyamoto, K., Kimura, Y. and Oda, K. 2016. A bacterium that degrades and assimilates poly(ethylene terephthalate). *Science*, 351(6278):1196–1199. doi:10.1126/science.aad6359.

7 Recent Bioengineering Advances in the Plastic Biodegradation and Future Challenges

7.1 INTRODUCTION

The extensive application of plastic in various sectors causes the accumulation of at least 100 million tons of plastics every year in nature. Polyethylene and polypropylene are widely employed for the fabrication of plastic bags, disposable containers, bottles, packaging materials, etc. and epitomize nearly 92% of the synthetic plastics being produced. The incessant disposal of crude plastic waste in open environments has been perceived as a significant environmental concern. Such a proliferating emanation of plastic might exert a severe toxicological effect on human health, which can be evinced by toxic chemical compounds, vectors of contaminants, and physical damage. Owing to meager plastic waste management techniques, such as mechanical and chemical recycling, landfills, incineration, etc., the implementation of microbial cultures in the remediation of plastic waste has transpired as an apt, economic, and eco-friendly strategy. Although there are few reports on the conversion of these polymers into environmentally friendly carbon compounds by some insects, bacteria, and fungi, still our current state of cognizance on the efficacy and prevailing mechanisms of plastic biodegradation seems inadequate.

Therefore, this chapter explores the application of genetically engineered microbes to achieve efficient degradation of synthetic plastic, and the enzymes involved in the process. In addition, the present chapter emphasizes the implementation of bioengineered microbes for the production of biopolymers. This chapter also highlights the potential of microbes to utilize plastic as a carbon source and provides a future outlook on the applications of system biology and computational biology in the field of plastic biodegradation.

7.2 BIOENGINEERING OF THE MICROBES FOR THE ENZYMES THAT DEGRADE NATURAL POLYMERS

Biopolymers are polymeric molecules having a covalent linkage of repetitive units of monomers (Kumari et al. 2021). On the basis of chemical structure, they can be classified into (i) polysaccharides, (ii) proteins, and (iii) polyesters (Aravamudhan et al. 2014). Competent microorganisms for biopolymer degradation have been recognized, and their potential was further explored in a plastic

DOI: 10.1201/9781003312086-7

degradation study. In recent synthetic plastic degradation studies, the polyethylene and polypropylene degradation by the microbes remains the cornerstone of the investigation (Park and Kim 2019). Microbial species with plastic degradation potential have been reported to belong to the genera *Bacillus*, *Aspergillus*, *Streptococcus*, *Klebsiella*, *Micrococcus*, *Staphylococcus*, and *Pseudomonas* (Skariyachan et al. 2017; Taghavi et al. 2021; Das and Kumar 2015). Numerous microbial species are reported to possess plastic degradation potential by utilizing it as a carbon and energy source, such as *Arthrobacter*, *Corynebacterium*, *Rhodococcus*, and *Streptomyces* (Jacquin et al. 2019). For instance, *Penicillium simplicissimum* YK was isolated and used in a polyethylene biodegradation study and the results showed potential of the isolated fungus to degrade polyethylene efficiently, depending on its growth phase (Yamada-Onodera et al. 2001). Similarly, Mehmood et al. (2016) studied the biodegradation of nanoparticles modified with photodegraded and non-photodegraded LDPE films, by isolating the bacterial isolates from solid waste dump sites. The prominent bacterial isolate showing the potential to degrade LDPE films was further characterized as *Stenotrophomonas pavanii*. In another study, *Alternaria alternata* FB1 was isolated from plastic waste-associated samples (Gao et al. 2022). Owing to its noteworthy potential for colonizing the polyethylene (PE) film, it was further used in the PE biodegradation study. The results revealed a sharp decline in the molecular weight of PE film by 95% and the enzymes involved in the degradation process were recognized as laccase and peroxidase. In a similar study, a novel bacterial isolate, *Ideonella sakaiensis* 201-F6, was evaluated to use polyethylene terephthalate (PET) as its prominent energy and carbon source (Shosuke et al. 2016). PET biodegradation by *I. sakaiensis* produces mono(2-hydroxyethyl) terephthalic acid (MHET) as a reaction intermediate. PET is hydrolyzed by extracellular PETase to yield MHET and terephthalate as the main products (Figure 7.1a). Janatunaim and Fibriani (2020) in silico, constructed recombinant *E. coli* BL21 (DE3) cells having genes for the plastics degrading enzymes (MHETase) from *I. sakaiensis* 201-F6, suggesting future applications in plastics degradation systems (Figure 7.1b). Following this, Espinosa et al. (2020) assessed the biodegradation of polyurethanes by isolating a bacterium from a site exposed to plastic waste. The strain was further characterized as *Pseudomonas* spp. and showed the noteworthy potential to use PU-diol solution, a polyurethane oligomer, and 2,4-diaminotoluene, a degradation intermediate of polyurethanes as the sole source of carbon and energy. Amines, alcohols, acids, aromatics, and other residues, such as ethylene glycol, 1,4-butanediol, adipic acid, methylenedianiline, and 2,4'-toluene diamine, are the degradation products formed from polyurethane (Figure 7.2). A conjunctive mix of polyesterases and urethane-degrading enzymes can be used to enhance degradation of the polyurethane waste (Liu et al. 2021a). Furthermore, engineering polyurethanes-specific binding domains can possibly enhance microbial degradation of solid polyurethanes (Liu et al. 2021a). In order to examine bacterial biodegradability of low-density polythene (LDPE), soil bacteria were isolated from a landfill (Biki et al. 2021). The bacterial isolates were characterized as *Ralstonia* spp. and *Bacillus* spp., which further showed a steady reduction in the weight of the LDPE sheet by about 39.2% and 18.9%, respectively. DSouza

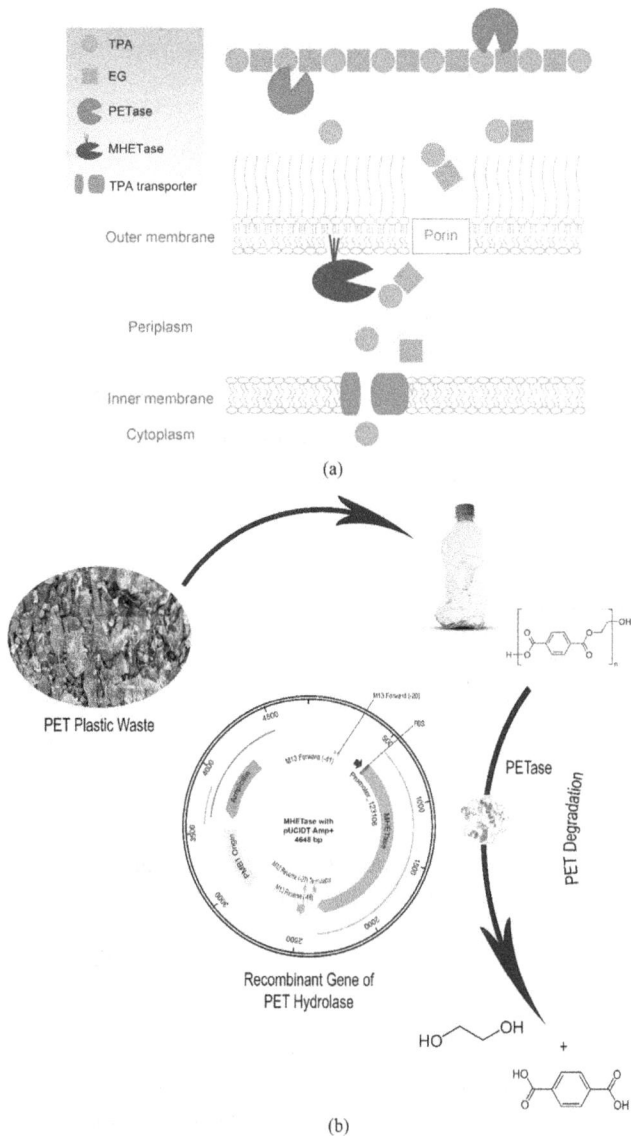

(a)

(b)

FIGURE 7.1 (a) PET metabolic pathway by *I. sakaiensis*. PET is hydrolyzed by extracellular PETase to yield MHET and terephthalate as the main products. The PET hydrolysis products are subsequently carried into the periplasmic region by an outer membrane protein called porin. MHETase (outer membrane lipoprotein) hydrolyzes MHET into terephthalate and ethylene glycol. A gene cluster found in *I. sakaiensis* is nearly identical to two TPA degradation gene clusters found in *Comamonas* spp. strain E6. Terephthalate upregulates the expression of this cluster in *I. sakaiensis*. TPA substantially increases the expression of this cluster in *I. sakaiensis*. Terephthalate is taken up into the cytoplasm by the terephthalate transporter and terephthalate-binding protein, and then incorporated into the tricarboxylic acid (TCA) cycle via protocatechuic acid (PCA). Via glyoxylic acid, ethylene glycol is metabolized to the TCA cycle. (b) Cloning of plastic-degrading recombinant enzymes (MHETase) in *E. coli* and applications in plastics degradation.

Source: (a) Taniguchi et al. (2019); (b) Janatunaim and Fibriani (2020).

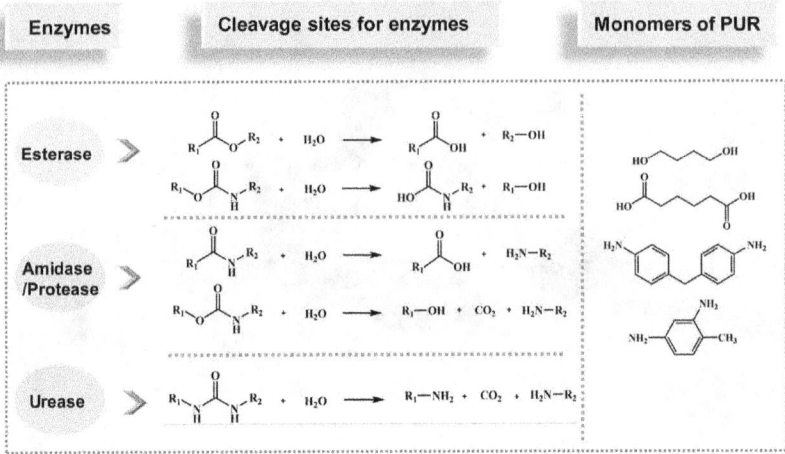

Enzymes	Cleavage sites for enzymes	Monomers of PUR

FIGURE 7.2 Enzymatic hydrolysis of chemical bonds in polyurethanes and released specific monomers.

Source: Liu et al. (2021a).

et al. (2021) studied biodegradation of LDPE using a fungal consortium encompassing *Aspergillus niger*, *A. flavus*, and *A. oryzae* and observed a diminished weight by 26.15%. In a recent study, *Brevundimonas* and *Sphingobacterium* spp. showed their PBAT/PLA polymer degradation potential under thermophilic conditions (Peng et al. 2022).

However, high throughput efficiency of synthetic plastic degradation by microbes is still not achieved due to certain bottlenecks associated with the use of microbial enzymes under industrial conditions, such as low catalytic efficiency, activity, and stability of enzymes. Application of synthetic microbial scavengers could improve pollutant removal efficiency and environment recovery. Therefore, research efforts are now being focused on the development of bioengineered microbes with improved efficiency for synthetic plastic degradation. Several strategies have been used for the bioengineering of microbes, including protein engineering, gene shuffling, and directed evolution, which have resulted in the development of enzymes with increased activities, vast specificity, and enhanced protein stability. For instance, in a study, the efficiency of polyethylene terephthalate (PET) plastic biodegradation was increased by designing DuraPETase-4M through protein engineering, which can be used to improve the efficiency of PET plastic biodegradation. The thermal stability, biodegradability of PET plastic, and binding ability between enzymes and PET plastic were increased by adding a pair of disulfide bonds (N233C/S282C), the key region flexibility adjustment (H214S) and protein surface electrostatic charge optimization, respectively (Liu et al. 2022). In another study, a modified *Y. lipolytica* yeast strain was designed to produce extracellularly cutinase from *Fusarium solani*. The modified strain subsequently showed PET degradation at 28°C during fermentation (Kosiorowska et al.

2022). In another study, the PETase enzyme was redesigned using photosynthetic microalga *Phaeodactylum tricornutum* and the results of the PET degradation study unraveled its activity against PET with nearly 80-fold enhanced turnover of low crystallinity polyethylene terephthalate glycol as compared to bottle PET (Moog et al. 2019).

Moreover, mutation design tools and computational strategies have been used in recent years to achieve higher PET degradation through bioengineered microbes. In a study, quantitative selection of the preferred mutations and incorporation of the sequence alignment were performed by generating a mutation design tool, Premuse. The results demonstrated about a 40-fold increase in amorphous PET degradation activity in comparison with the wild type (Meng et al. 2021). Another study based on a systematic clustering analysis combined with a computational strategy enabled the modification of a variant, DuraPETase, from *Ideonella sakaiensis*. The results showed an increase in apparent melting temperature by 31°C along with an escalated degradation towards semicrystalline PET films at mild temperatures (Cui et al. 2021).

7.3 BIOENGINEERING OF THE MICROBES FOR THE PRODUCTION OF BIOPOLYMERS

Biopolymers are regarded as renewable, biodegradable, and sustainable alternatives to petrochemical polymers, including synthetic plastic. These are known to be produced from various kinds of microorganisms such as bacteria, algae, and fungi. For instance, *Methylobacterium* spp. ISTM1, and *Pseudomonas umsongensis* GO16 are reported to produce polyhydroxyalkanoates (PHA), while *Providencia* spp., is known for its potential for polyhydroxybutyrate (PHB) synthesis. Besides, some other compounds involved in the synthesis of biopolymers are also known to be synthesized by microorganisms. For example, vanillin, a key-intermediate of bio-based polymers and itaconic acid, an additive used in biopolymer synthesis, is reported to be produced from a modified strain of *Escherichia coli* and *Ustilago* spp., respectively (Sadler and Wallace 2021; Nascimento et al. 2022). However, microbial engineering has widened the dimensions of green biological chemistry by escalating the efficiency of biopolymer production (Figure 7.3).

Sustainable yields of biopolymers can be maintained through microbial synthesis, which can be further magnified through genetic engineering approaches designed for constructing engineered microorganisms with high efficiency and capable to produce biopolymers like polyhydroxyalkanoates (PHAs), polylactic acid (PLA), glycosaminoglycan, etc. (Verma et al. 2020; Liu et al. 2021b). PHAs are the most commonly used biopolymers owing to their plastic-like characteristics, biodegradability, and eco-friendly nature. Their chemical structure is composed of polyesters of hydroxyalkanoates, which can be synthesized from renewable compounds such as organic waste by microbial activity (Guleria et al. 2022). Bacterial species can be reconstructed using synthetic biology to generate low-cost intracellular PHAs (Figure 7.4). In a study, low-cost production of PHA was achieved through redesigning of *Halomonas* bacteria (Chen et al. 2016). Kamravamanesh et al. (2018) used the random mutagenesis method by applying UV light as a mutagen to reconstruct

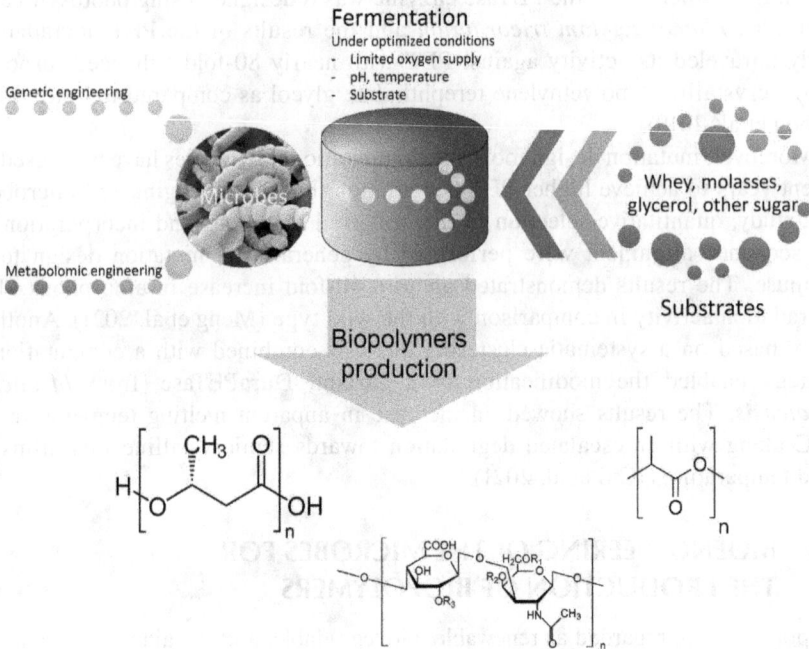

FIGURE 7.3 Schematic representation of the strategies used to enhance the sustainable production of biopolymers.

Synechocystis spp. PCC 6714. The resultant mutant MT_a24 demonstrated rapid growth and more than 2.5-fold higher PHB productivity compared to wild-type. Koch et al. (2020) developed a strategy for boosting PHB production by redirecting carbon from fatty acids, which resulted in the over-expression of the gene encoding FabG protein. In a study, the *H. bluephagenesis* strain was engineered by using the optimized expression of aldehyde dehydrogenase (AldD$_{Hb}$). The strain was redesigned to over-express alcohol dehydrogenases (AdhP). The results demonstrated the potential of the reconstructed strain to synthesize 3-Hydroxypropionic acid (Jiang et al. 2021). In another study, a recombinant strain of *C. necator* DSM 545 was developed by co-expressing the glucodextranase *Gld* and α-amylase *amyZ* from *Arthrobacter globiformis* I42 and *Zunongwangia profunda* SM-A8, respectively (Brojanigo et al. 2022). Recently, recombinant *E. coli* was developed by cloning novel *phaCABp* genes from *Propylenella binzhouense* L72^T (Figure 7.4). The results indicated a fairly high poly(3-hydroxybutyrate-co-3-hydroxyvalerate) (PHBV) yield of about 1.06 g L^{-1} (Meng et al. 2022). Polymer from the developed recombinant strain had better performance and thermostability. Wang et al. (2014) used a novel strategy of developing a filamentary recombinant *E. coli* by over-expression of the *sulA* gene to enhance PHA accumulation. This led to the inhibition of cell division with subsequent enlargement of *E. coli*, providing more area for PHA accumulation. The results revealed a higher amount of poly(3-hydroxybutyrate) (PHB) in the recombinant strain compared to its control.

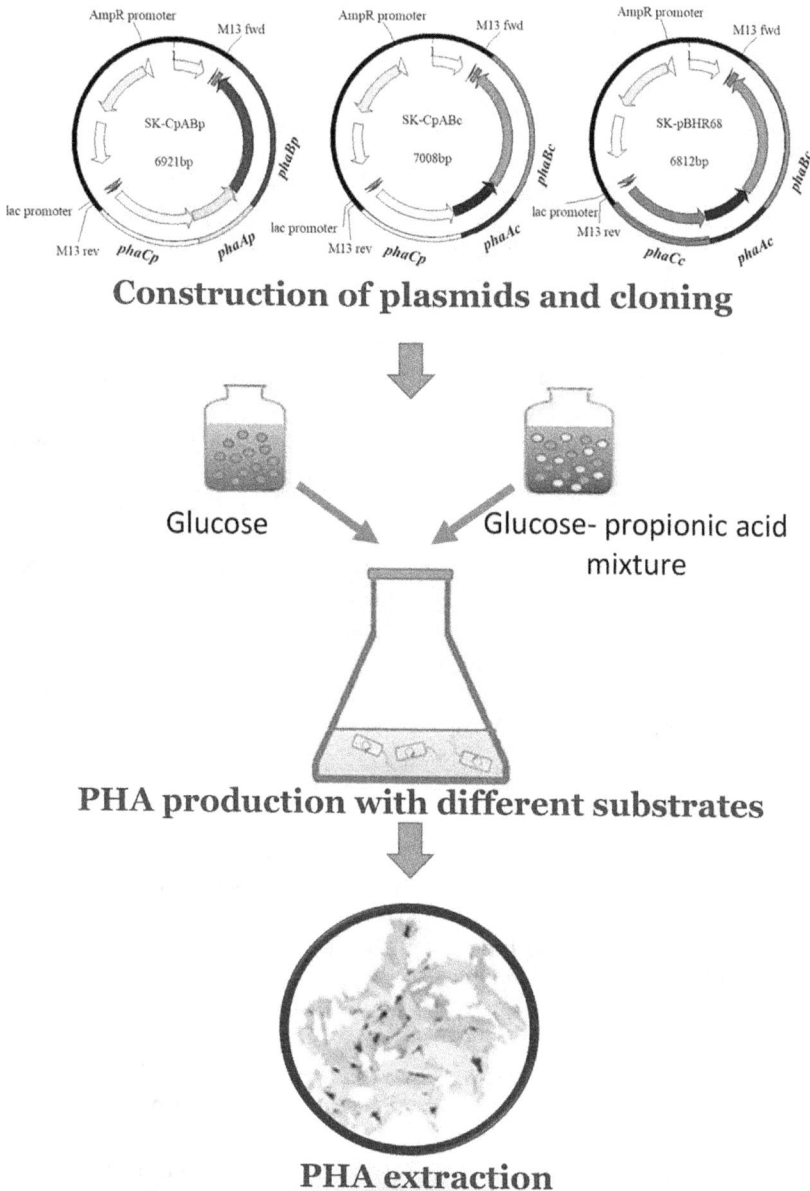

Construction of plasmids and cloning

Glucose

Glucose- propionic acid mixture

PHA production with different substrates

PHA extraction

FIGURE 7.4 Metabolic engineering for production of poly(3-hydroxybutyrate-co-3-hydroxyvalerate) (PHBV) a polyhydroxyalkanoates (PHAs) family of biopolymer from glucose and propionic acid using recombinant *Escherichia coli*.

Source: Meng et al. (2022).

7.4   FUTURE RESEARCH DIRECTIONS

Recent studies on the application of microbial engineering in plastic waste management have given a deep insight into the potent plastic degrading microbial strains and their molecular mechanisms. However, there are still some bottlenecks associated with the biodegradation of synthetic plastic that need to be addressed.

1. Recombinant strains developed in several plastic degradation studies showed low performance is not suited for large-scale application. Discovery of more potential plastic degrading enzymes and evaluation of their crystal structure and degradation mechanisms could pave the way for industrial application.
2. Confluent application of omics, systems biology, and bioinformatics such as genome editing, systems, and synthetic biology, and real-time biopolymer characterization methods in recombinant biology could produce the microbes with desired traits with superior performance in plastic degradation and ensure good quality of synthesized biopolymers.
3. Advancement in computational approaches, such as machine learning and artificial intelligence, along with emerging functional metagenomics, metatranscriptomics and metaproteomics could make a breakthrough in unraveling protein structure, plastic-microbe interactions, and providing a blueprint of protein engineering.
4. Research directions should be directed towards development of biodegradable plastics with facile downstream processing techniques, which could ensure cost-effectiveness.

7.5   SUMMARY

Polyethylene and polypropylene are widely employed for the fabrication of plastic bags, disposable containers, bottles, packaging materials, etc. There are few reports on the conversion of these polymers into environmentally friendly carbon compounds by some insects, bacteria, and fungi. The genetic engineering of the microbes can be used to achieve efficient degradation of synthetic plastic. Some of the microbial species having plastic degradation potential reported in the recent studies belong to the following genera: *Bacillus*, *Aspergillus*, *Penicillium*, *Streptococcus*, *Klebsiella*, *Micrococcus*, *Staphylococcus*, and *Pseudomonas*.

Protein engineering, gene shuffling, and directed evolution have all been used for the bioengineering of microbes, resulting in the generation of enzymes with greater activity, broad specificity, and improved protein stability. High throughput efficiency of synthetic plastic degradation by microbes is still not achieved due to bottlenecks associated with the use of microbial enzymes under industrial conditions. The development of bioengineered microorganisms with higher efficiency for synthetic plastic biodegradation is presently the focus of research.

Alternatives to petrochemical polymers include biopolymers, which are renewable, sustainable, and biodegradable. These biopolymers are known to be produced by microbes, including bacteria, algae, and fungi. By increasing the efficiency of biopolymer manufacturing, microbial engineering has expanded the scope of green biological chemistry. Recombinant DNA technology and metabolic engineering can be used to enhance the production of biodegradable polyesters by the bacteria. For example, Wang and coworkers created a filamentary recombinant *E. coli* in 2014 by over-expressing the *sulA* gene in order to increase PHA accumulation. Recently, Brojanigo and coworkers developed a recombinant strain of *C. necator* DSM 545 by co-expressing the glucodextranase *Gld* and α-amylase *amyZ* from *Arthrobacter globiformis* I42 and *Zunongwangia profunda* SM-A8, respectively. At higher processing temperatures, the polymer from the recombinant strain performed better.

REFERENCES

Aravamudhan, A., Ramos, D. M., Nada, A. A. and Kumbar, S. G. 2014. Natural Polymers: Polysaccharides and Their Derivatives for Biomedical Applications. *Natural and Synthetic Biomedical Polymers*, 67–89. doi: 10.1016/B978-0-12-396983-5.00004-1.

Biki, S. P., Mahmud, S., Akhter, S., Rahman, M. J., Rix, J. J., Al Bachchu, M. A. and Ahmed, M. 2021. Polyethylene degradation by *Ralstonia* sp. strain SKM2 and Bacillus sp. strain SM1 isolated from land fill soil site. *Environmental Technology and Innovation*, 22:101495. doi: 10.1016/j.eti.2021.101495.

Brojanigo, S., Gronchi, N., Cazzorla, T., Wong, T. S., Basaglia, M., Favaro, L. and Casella, S. 2022. Engineering Cupriavidus necator DSM 545 for the one-step conversion of starchy waste into polyhydroxyalkanoates. *Bioresource Technology*, 347:126383 doi: 10.1016/j.biortech.2021.126383.

Chen, G.-Q., Jiang, X.-R. and Guo, Y. 2016. Synthetic biology of microbes synthesizing polyhydroxyalkanoates (PHA). *Synthetic and Systems Biotechnology*, 1(4):236–242. doi: 10.1016/j.synbio.2016.09.006.

Cui, Y., Chen, Y., Liu, X., Dong, S., Tian, Y., Qiao, Y., Mitra, R., Han, J., Li, C., Han, X., Liu, W., Chen, Q., Wei, W., Wang, X., Du, W., Tang, S., Xiang, H., Liu, H., Liang, Y., . . . Wu, B. 2021. Computational redesign of a PETase for plastic biodegradation under ambient condition by the GRAPE strategy. *ACS Catalysis*, 11(3):1340–1350. doi: 10.1021/acscatal.0c05126.

Das, M. P. and Kumar, S. 2015. An approach to low-density polyethylene biodegradation by Bacillus amyloliquefaciens. *3 Biotech*, 5(1):81–86. doi: 10.1007/s13205-014-0205-1.

DSouza, G. C., Sheriff, R. S., Ullanat, V., Shrikrishna, A., Joshi, A. V., Hiremath, L. and Entoori, K. 2021. Fungal biodegradation of low-density polyethylene using consortium of Aspergillus species under controlled conditions. *Heliyon*, 7(5):e07008.

Espinosa, M. J. C., Blanco, A. C., Schmidgall, T., Atanasoff-Kardjalieff, A. K., Kappelmeyer, U., Tischler, D., Pieper, D. H., Heipieper, H. J. and Eberlein, C. 2020. Toward biorecycling: Isolation of a soil bacterium that grows on a polyurethane oligomer and monomer. In *Frontiers in Microbiology* (Vol. 11). www.frontiersin.org/article/10.3389/fmicb.2020.00404.

Gao, R., Liu, R. and Sun, C. 2022. A marine fungus Alternaria alternata FB1 efficiently degrades polyethylene. *Journal of Hazardous Materials*, 431:128617. doi: 10.1016/j.jhazmat.2022.128617.

Guleria, S., Singh, H., Sharma, V., Bhardwaj, N., Arya, S. K., Puri, S. and Khatri, M. 2022. Polyhydroxyalkanoates production from domestic waste feedstock: A sustainable approach towards bio-economy. *Journal of Cleaner Production*, 340:130661. doi: 10.1016/j.jclepro.2022.130661.

Jacquin, J., Cheng, J., Odobel, C., Pandin, C., Conan, P., Pujo-Pay, M., Barbe, V., Meistertzheim, A.-L. and Ghiglione, J.-F. 2019. Microbial ecotoxicology of marine plastic debris: A review on colonization and biodegradation by the "plastisphere". In *Frontiers in Microbiology* (Vol. 10). www.frontiersin.org/article/10.3389/fmicb.2019.00865.

Janatunaim, R. Z. and Fibriani, A. 2020. Construction and cloning of plastic-degrading recombinant enzymes (MHETase). *Recent Patents on Biotechnology*, 14(3):229–234. http://dx.doi.org/10.2174/1872208314666200311104541.

Jiang, X.-R., Yan, X., Yu, L.-P., Liu, X.-Y. and Chen, G.-Q. 2021. Hyperproduction of 3-hydroxypropionate by *Halomonas bluephagenesis*. *Nature Communications*, 12(1):1513. doi: 10.1038/s41467-021-21632-3.

Kamravamanesh, D., Kovacs, T., Pflügl, S., Druzhinina, I., Kroll, P., Lackner, M. and Herwig, C. 2018. Increased poly-β-hydroxybutyrate production from carbon dioxide in randomly mutated cells of cyanobacterial strain Synechocystis sp. PCC 6714: Mutant generation and characterization. *Bioresource Technology*, 266:34–44.

Koch, M., Berendzen, K. W. and Forchhammer, K. 2020. On the role and production of poly-hydroxybutyrate (PHB) in the cyanobacterium synechocystis sp. PCC 6803. *Life*, 10(4). doi: 10.3390/life10040047.

Kosiorowska, K. E., Biniarz, P., Dobrowolski, A., Leluk, K. and Mirończuk, A. M. 2022. Metabolic engineering of *Yarrowia lipolytica* for poly(ethylene terephthalate) degradation. *Science of The Total Environment*, 831:154841. doi: 10.1016/j.scitotenv.2022.154841.

Kumari, P., Lal, S. and Singhal, A. 2021. Advanced applications of green materials in catalysis applications. *Applications of Advanced Green Materials*, 545–571. doi: 10.1016/B978-0-12-820484-9.00022-2.

Liu, H., Wei, L., Ba, L., Yuan, Q. and Liu, Y. 2021b. Biopolymer production in microbiology by application of metabolic engineering. *Polymer Bulletin*. doi: 10.1007/s00289-021-03820-9.

Liu, J., He, J., Xue, R., Xu, B., Qian, X., Xin, F., Blank, L. M., Zhou, J., Wei, R., Dong, W. and Jiang, M. (2021a). Biodegradation and up-cycling of polyurethanes: Progress, challenges, and prospects. *Biotechnology Advances*, 48:107730. doi: 10.1016/j.biotechadv.2021.107730.

Liu, Y., Liu, Z., Guo, Z., Yan, T., Jin, C. and Wu, J. 2022. Enhancement of the degradation capacity of IsPETase for PET plastic degradation by protein engineering. *Science of the Total Environment*, 834:154947. doi: 10.1016/j.scitotenv.2022.154947.

Mehmood, C. T., Qazi, I. A., Hashmi, I., Bhargava, S. and Deepa, S. 2016. Biodegradation of low density polyethylene (LDPE) modified with dye sensitized titania and starch blend using Stenotrophomonas pavanii. *International Biodeterioration and Biodegradation*, 113:276–286. doi: 10.1016/j.ibiod.2016.01.025.

Meng, D., Miao, C., Liu, Y., Wang, F., Chen, L., Huang, Z., Fan, X., Gu, P. and Li, Q. 2022. Metabolic engineering for biosynthesis of poly(3-hydroxybutyrate-co-3-hydroxyvalerate) from glucose and propionic acid in recombinant Escherichia coli. *Bioresource Technology*, 348:126786. doi: 10.1016/j.biortech.2022.126786.

Meng, X., Yang, L., Liu, H., Li, Q., Xu, G., Zhang, Y., Guan, F., Zhang, Y., Zhang, W., Wu, N. and Tian, J. 2021. Protein engineering of stable IsPETase for PET plastic degradation by Premuse. *International Journal of Biological Macromolecules*, 180:667–676. doi: 10.1016/j.ijbiomac.2021.03.058.

Moog, D., Schmitt, J., Senger, J., Zarzycki, J., Rexer, K.-H., Linne, U., Erb, T. and Maier, U. G. 2019. Using a marine microalga as a chassis for polyethylene terephthalate (PET) degradation. *Microbial Cell Factories*, 18(1):171. doi: 10.1186/s12934-019-1220-z.

Nascimento, M. F., Marques, N., Correia, J., Faria, N. T., Mira, N. P. and Ferreira, F. C. 2022. Integrated perspective on microbe-based production of itaconic acid: From metabolic and strain engineering to upstream and downstream strategies. *Process Biochemistry*, 117:53–67. doi: 10.1016/j.procbio.2022.03.020

Park, S. Y. and Kim, C. G. 2019. Biodegradation of micro-polyethylene particles by bacterial colonization of a mixed microbial consortium isolated from a landfill site. *Chemosphere*, 222:527–533. doi: 10.1016/j.chemosphere.2019.01.159.

Peng, W., Wang, Z., Shu, Y., Lü, F., Zhang, H., Shao, L. and He, P. 2022. Fate of a biobased polymer via high-solid anaerobic co-digestion with food waste and following aerobic treatment: Insights on changes of polymer physicochemical properties and the role of microbial and fungal communities. *Bioresource Technology*, 343:126079. doi: 10.1016/j. biortech.2021.126079.

Sadler, J. C. and Wallace, S. 2021. Microbial synthesis of vanillin from waste poly(ethylene terephthalate). *Green Chemistry*, 23(13):4665–4672. doi: 10.1039/D1GC00931A.

Shosuke, Y., Kazumi, H., Toshihiko, T., Ikuo, T., Hironao, Y., Yasuhito, M., Kiyotsuna, T., Kenji, M., Yoshiharu, K. and Kohei, O. (2016). A bacterium that degrades and assimilates poly(ethylene terephthalate). *Science*, 351(6278):1196–1199. doi: 10.1126/science.aad6359.

Skariyachan, S., Setlur, A. S., Naik, S. Y., Naik, A. A., Usharani, M. and Vasist, K. S. 2017. Enhanced biodegradation of low and high-density polyethylene by novel bacterial consortia formulated from plastic-contaminated cow dung under thermophilic conditions. *Environmental Science and Pollution Research*, 24(9):8443–8457. doi: 10.1007/s11356-017-8537-0.

Taghavi, N., Singhal, N., Zhuang, W.-Q. and Baroutian, S. 2021. Degradation of plastic waste using stimulated and naturally occurring microbial strains. *Chemosphere*, 263:127975. doi: 10.1016/j.chemosphere.2020.127975.

Taniguchi, I., Yoshida, S., Hiraga, K., Miyamoto, K., Kimura, Y. and Oda, K. 2019. Biodegradation of PET: Current status and application aspects. *Acs Catalysis*, 9(5):4089–4105. doi: 10.1021/acscatal.8b05171.

Verma, M. L., Kumar, S., Jeslin, J. and Dubey, N. K. 2020. Microbial *Production of Biopolymers with Potential Biotechnological Applications* (K. Pal, I. Banerjee, P. Sarkar, D. Kim, W.-P. Deng, N. K. Dubey, and K. B. T.-B.-B. F. Majumder (eds.); pp. 105–137). London: Elsevier. doi: 10.1016/B978-0-12-816897-4.00005-9.

Wang, Y., Wu, H., Jiang, X. and Chen, G.-Q. 2014. Engineering Escherichia coli for enhanced production of poly(3-hydroxybutyrate-co-4-hydroxybutyrate) in larger cellular space. *Metabolic Engineering*, 25:183–193. doi: 10.1016/j.ymben.2014.07.010.

Yamada-Onodera, K., Mukumoto, H., Katsuyaya, Y., Saiganji, A. and Tani, Y. 2001. Degradation of polyethylene by a fungus, Penicillium simplicissimum YK. *Polymer Degradation and Stability*, 72(2):323–327. doi: 10.1016/S0141-3910(01)00027-1.

8 Biopolymers as an Alternative to the Conventional Plastics

Conventional types of plastics have proved detrimental to Earth's ecosystems and hazardous to living organisms and human health. With a high degree of resistance to biodegradation and ubiquitous nature, the conventional plastics, fragmented into microplastics and nanoplastics, are there to prevail on the planet. Their intensive utilization in almost all socioeconomic systems followed by mismanagement of utilized products and wastes is leaving behind a trail of unprecedented problems which could be difficult to manage in the future. Human society in the contemporary world is so used to the plastic items that it would be pretty difficult to halt their production. What is readily possible is bringing the alternatives to conventional plastics into use. Since resistance of the conventional plastics to biodegradability has been a major cause for their prevalence in nature for decades, their biodegradable alternatives—or biopolymers—can greatly help get rid of the microplastic pollution in due course of time.

8.1 WHAT ARE BIOPOLYMERS?

Biopolymers are the polymers derived from natural sources, biosynthesized from living organisms or chemically synthesized using biological material. These are composed of monomeric units covalently bonded to form larger molecules. Since the biopolymers are biologically derived, they are biodegradable, unlike most of the petroleum-based conventional plastics.

8.1.1 WHY ARE BIOPOLYMERS NEEDED?

Detrimental impacts of global proportions throughout the life cycle of petroleum-based, mostly non-biodegradable and ubiquitous conventional plastics, such as microplastic pollution in marine and terrestrial ecosystems, carbon emissions and toxins released upon incineration, and associated human health hazards, underline the necessity for biopolymers derived from natural resources. Microplastic emissions and their serious implications for all aspects of the planet's life, including for humankind, cannot be totally adopted despite the implementation of most appropriate measures. Biopolymers as an alternative to the conventional plastics, in fact, offer the most promising solution to this indefensible problem that our contemporary world is facing.

DOI: 10.1201/9781003312086-8

8.2 BIOPOLYMERS

Most of the living world is based on polymers. They inherently occur in animals (proteins, nucleic acids, hydrocarbons, etc.), plants (starch, cellulose, oils, etc.), and lower organisms (Šprajcar et al. 2012). The microbe-derived polymers include polyhydroxyalkanoate.

8.2.1 BIO-BASED VS PETROCHEMICAL-BASED POLYMERS

A few decades ago, almost all polymers were petroleum-based. For some years now we can witness many types of bio-based polymers in use for various purposes. The bio-based polymers are manufactured by using biomass as raw material, that is, by using renewable resources, while petrochemical-based ones are derived from non-renewable and exhaustible resources. Production of biopolymers would have considerably less negative impact on the climate compared to that of the petrochemical-based ones. The bio-based products are carbon-neutral or offset carbon emissions. The petrochemical-based polymers, on the other hand, contribute to the production of greenhouse gases and, thus, to climate change. The bio-based polymers are mostly biodegradable and compostable and are of multiple uses—including as a valuable source for soil fertility after their use. Biopolymers—in essence—are part of the sustainability processes.

8.2.2 PROTEIN AS A BIOPOLYMER

Proteins are biopolymers with some 20 amino acids linked with each other through amide bonds. Certain specificities and suitability of proteins are helpful in developing them into biopolymers as an alternative to traditional fossil fuel-based polymers. Polymer reinforcement techniques like development of blends, chemical block copolymerization, and modification of existing protein material are usable for developing protein-based polymers often used for food packaging and in the health sector (Gupta and Nayak 2014). Some of the typical protein polymers are plant proteins (e.g., zein, gluten, etc.) and animal proteins (e.g., silk, collagen, actin, fibrin, keratin, elastin, etc.).

8.2.2.1 Collagen and Gelatin

Collagen, the most important structural protein in the vertebrates and most abundant among mammals, is a biopolymer of versatile applications. It has high tensile strength which is attributable to its mechanical structure. As a readily available protein-based polymer, collagen is widely applicable in medical sciences. Gelatin is a form of protein biopolymer derived from collagen through controlled hydrolysis. There are two types of gelatin—type A and type B—that can be formed by collagen pretreatment with acid or alkali. Gelatin comprises large numbers of glycine, proline, and 4-hydroxyproline residues.

8.2.2.2 Silk

Silk is a protein-rich biopolymer derived from various silk worm species, for example the mulberry worm *Bombyx mori*. The insoluble and fibrous protein in the composition of silk makes the product somewhat adhesive. In its tensile strength, the silk is weaker than collagen. Anticoagulation properties and platelet adhesion associated with silk and its role in *in vitro* stem cell prolification makes the silk fibroin highly applicable in medical sciences.

8.2.3 POLYSACCHARIDES

Among the most important and popular bio-based polymers are those made from polysaccharides. Starch and cellulose are the main constituents of polysaccharide-based bioplastics. Chitin is the other, comparatively less popular, polysaccharide occurring mainly in insect shells. Some examples of bio-based polymers originating from polysaccharides are cellophane and cellulose acetate made from cellulose, thermoplastic starch made of starch, and microbe-derived thermoplastic polylactic acid (PLA).

8.2.3.1 Starch

Starch is one of the most abundant and interesting raw materials for producing biopolymers. Native starches, however, show poor mechanical properties and thermal stability, and high humidity absorption capacity. Owing to these distinguishable properties, the starches are blended with other materials to produce products exhibiting desirable qualities, such as high biodegradability, enhanced mechanical integrity, and declined hydrophilic properties. Starches are blended with plasticizers exhibiting low molecular mass (e.g., ethylene glycol, sorbitol, glycerol, etc.). Such a blending helps optimize the properties of the starch-plasticizer blended product (Tang and Alavi 2011; Encalada et al. 2018). Significant improvements in the thermoplastic starch (TPS) properties can be made by blending them with other polymers, both natural and synthetic, such as cellulose, maize protein zein, polyvinyl alcohol (PVA), polyesters, polyurethanes (PUR), etc. TPS blending with biodegradable polymers (e.g., PVA and PLA) and with natural fibers, like wood pulp and hemp, provide usable products with 100% renewability and biodegradability. Such types of plastics in use would be ecologically beneficial and help resolve many problems arising out of the use of conventional plastic materials.

8.2.3.2 Cellulose

Cellulose, a structural polysaccharide in plants, is almost an inexhaustible polymer usable in the production of eco-friendly and sustainable biopolymers. One of the most abundantly used polymers, cellulose has many unique properties making it an environmentally safe, ecologically promising, and socioeconomically useful alternative to the detrimental conventional polymers. Cellulose and its derivatives are widely used in the food packaging industry (Liu et al. 2021). When microplastic pollution emerging from conventional plastics is becoming a nuisance for the whole world, the most readily available natural polymer might be utilized as an alternative in wide applications.

8.2.3.3 Chitin/Chitosan

Chitin (β-(1–4)-poly-N-acetyl-D-glucosamine) is the most abundantly occurring aminopolysaccharide polymer and the second most abundantly found polysaccharide in nature, next to cellulose. This polymer imparts strength to the exoskeletons of crustaceans, insects, and fungi cell walls. Chitosan is the well-known derivative of chitin obtainable through chemical or enzymatic deacetylation of chitin. Chitin biosynthesis can be catalyzed by the widely occurring enzyme in nature among the chitin-synthesizing organisms, chitin synthase. Chitinase, an enzyme widely distributed in nature, is responsible for chitin's biodegradation. Chitosan, having biocompatibility with chitin, is also biodegradable and non-toxic (Elieh-Ali-Komi and Hamblin 2016). Both chitin and chitosan are highly applicable in medical sciences, like in wound healing, tissue engineering, drug and gene delivery, and stem cell technology (Azuma et al. 2014). The biopolymers can also be readily processed into a variety of items, such as beads, hydrogels, nanofibers, scaffolds, microparticles, nanoparticles, sponges, etc. and brought into a range of socioeconomic applications.

8.2.4 POLYHYDROXYALKANOATES

Polyhydroxyalkanoates (PHAs) are the natural polyester polymers synthesized by bacteria quite akin to conventional petrochemical polymers in their physical properties. They can be produced using agro-industrial waste streams and can be appropriately tailored in various industrial applications. As renewable and fully biodegradable bio-based polymers, the PHAs represent a potential platform for bioplastics, bioresins, and biocomposite materials. Microalgae have been found to be effective microorganisms in PHA synthesis. As photoautotrophs, microalgae can function in the presence of light and thus would help reduce the costs of PHA production. Bacteria used for synthesis can be genetically modified to enhance biopolymer production.

8.3 BIODEGRADABILITY AND COMPOSTABILITY

Increasing human consciousness about the quest for sustainability creates a demand for the production and utilization of products that are biodegradable as well as compostable. These two properties of the polymers being overwhelmingly used in daily life would phenomenally help in ameliorating the ecological processes leading to ecological sustainability—a precondition for attaining socioeconomic sustainability (Singh 2019, 2020). Ecological sustainability is also the basis of climate regulation and a phenomenon for creating a climate smart planet (Singh 2020).

What is biodegradable and what is compostable? The two terms appear to be indistinguishable, but they are not. There are some scientific and legal differences in the concepts about the biodegradable and compostable products in use. A biodegradable product is not always compostable, but a compostable one is always biodegradable. A compostable material is a subset of a biodegradable material, in essence.

Biodegradability indicates that a material will fully degrade within a specified period of its disposal. Claiming a product being "biodegradable" without reliable scientific evidence is illegal. Article 260.8(c) of the Green Guides issued by the Federal Trade Commission (FTC) says: "It is deceptive to make an unqualified degradable claim for items entering the solid waste system if the items do not completely decompose within one year after customary disposal".

Biodegradable plastics, according to the US Environmental Protection Agency (EPA), include the kind of natural polymers that are helpful in getting degraded into smaller components when exposed to an appropriate environment. In order to set up standards to measure biodegradation under aerobic conditions, the American Society for Testing and Materials (ASTM) created ASTM D5338. The ASTM D6691 is meant for measuring biodegradability under marine conditions. In the USA, the biodegradability standards defined by ASTM are implemented by the FTC. A biodegradable plastic undergoes biodegradation in soil or in water. Unless it is labeled as only marine biodegradable, it is usable for composting at a commercial/industrial scale.

A compostable product, according to the EPA, is the one that meets ASTM D6400. A compostable type of polymer is supposed to be biodegradable into "soil conditioning material". The composting processes are carried out under favorable conditions involving specific microorganisms, moisture, and suitable temperature. Standards for commercial or industrial compostability are set out under ASTM D6400 and D6868. These standards pertain to the compostable waste products to be used at industrial or commercial scale, not to the products usable for composting at home.

While all the manufactured products defined as biodegradable and compostable are ultimately recyclable, all the recyclable ones need not be essentially compostable. The biodegradable and compostable processes relating to industrial products have to accomplish degradation within a defined timeframe, while there is no such period affixed with the recyclable material; biodegradable and compostable products would return to their pool in nature as per the inherent properties of the material. Biodegradability, after all, is the central characteristic of a product that determines its role and fate in nature.

8.4   MICROBES AS THE SOURCE OF BIOPOLYMERS

Microorganisms play an important role in producing several polymers. Exopolysaccharides (EPS) produced extracellularly and polyhydroxyalkanoates (PHAs) produced intracellularly are the two naturally produced biopolymers assuming preference over other biopolymers and can substitute for plant- and fossil fuel-based polymers. These important biopolymers have been successfully produced using a variety of microorganisms: bacteria, actinomycetes, fungi, and algae (Angelina and Vijayendra 2015). Microbial biomass can be used as the material to produce biopolymers at industrial and commercial levels. Microbial cellulose is a more promising biopolymer than plant cellulose (Cottet et al. 2020). One of the benefits of microbe-derived biopolymers is that their commercial production is not limited by factors like crop failure, climate conditions, or ocean pollution. Some of the microbe-derived biopolymers may be more expensive than the plant-based ones. However, biopolymers obtainable from fungal biomass may be far less expensive because of abundance of the biomass material left as residues of fungal cells and mycelia during some industrial processes applied in the brewing industry (Kadimaliev et al. 2015; Peltzer et al. 2018) and other biotechnology-based industries (Guimaraes et al. 2015).

Microbial biomass originating from fungi, especially the yeast and mycelium biomass, has enormous potential for use in various technologies. For example, the biopolymers developed from yeast biomass are used for encapsulation of bioactive

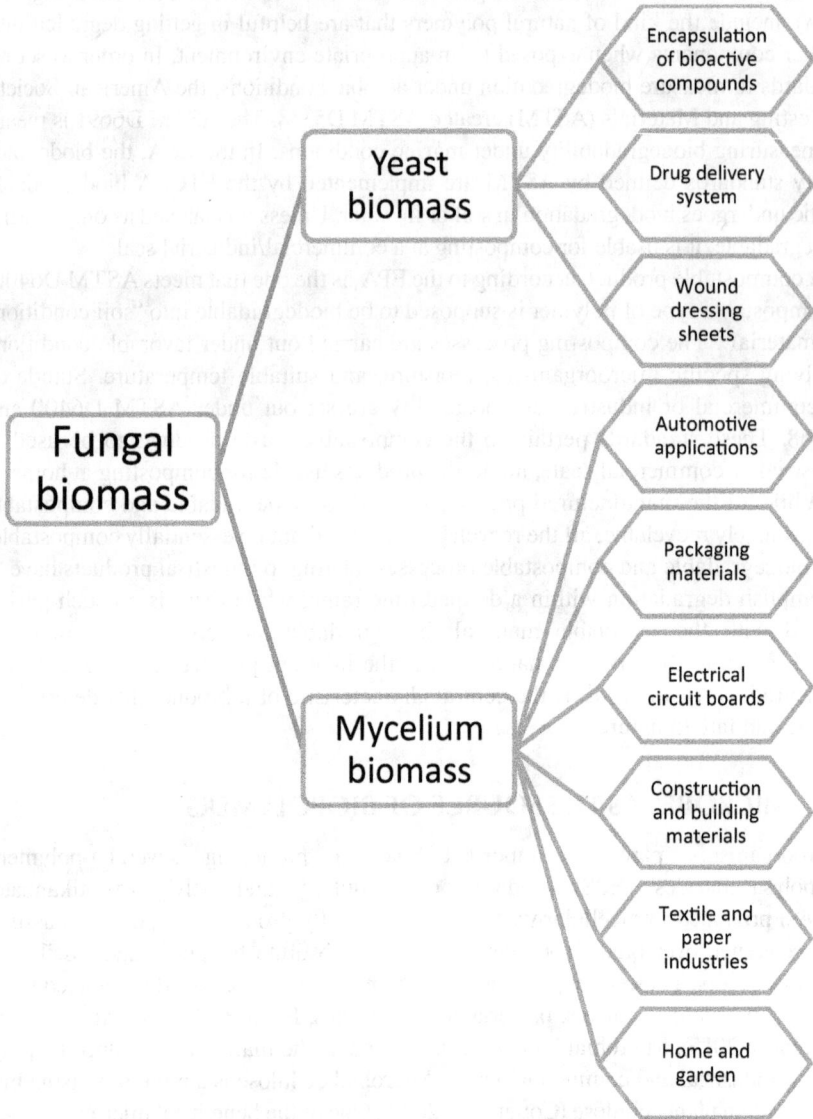

FIGURE 8.1   Different uses of fungal biomass.

compounds (Mokhtari et al. 2017), wound dressing sheets, and drug delivery systems (e.g., use of bio-inspired yeast microcapsules in the delivery of charged nanoparticles) (Sabu et al. 2019). The mycelium biomass is usable in developing biopolymers of potential applications in various technologies such as in automotive applications, electrical circuit boards, construction and building materials, home and garden, and packaging materials (Cerimi et al. 2019) (Figure 8.1). Kefir grains and bacterial cellulose are of wide industrial applications for producing a variety of biodegradable biopolymers usable for various purposes.

Kefir grains, supposedly originating in the Caucasus Mountains, are a symbiotic culture comprising bacteria and yeast. The grains contain lactic acid bacteria (LAB) and acetic acid bacteria (AAB). The main polysaccharide is kefiran with about equal amounts of glucose and galactose and produced by *Lactobacillus kefiranofaciens* (Radhouani et al. 2018). The kefiran and dextran-based materials are of crucial applications in biomedical and food packaging areas. In the biomedical field these biopolymers are especially of significance in tissue engineering (Radhouani et al. 2019), skin regeneration (Radhouani et al. 2018), and drug delivery systems (Radhouani et al. 2019). In addition, the polymer is usable in active food packaging (Vijayendra and Shamala 2014). Kefiran extract carries the strongest reducing power and superoxide radical scavenging, over hyaluronic acid (HA). Capacity of scavenging nitric oxide radical is another property associated with this EPS. Kefiran demonstrated no cytotoxic effects and has been found to possess the ability to improve cellular functions of human adipose stem cells (hASCs) (Figure 8.2). This EPS serves as a grand scavenger for reactive oxygen and nitrogen species. It can serve as a superb candidate for the promotion of tissue repair and regeneration (Radhouani et al. 2018).

Microbial cellulose, especially bacterial cellulose, has multiple properties of a biopolymer with several applications in many areas and sectors, notably foods, pharmaceuticals, medical sciences, engineering, and the environment as noted in Table 8.1.

Bacillus subtilis is used to ferment glutamic acid forming gamma-poly-glutamic acid (gamma-PGA), a multi-functional and biodegradable polymer. Fungal bio-based

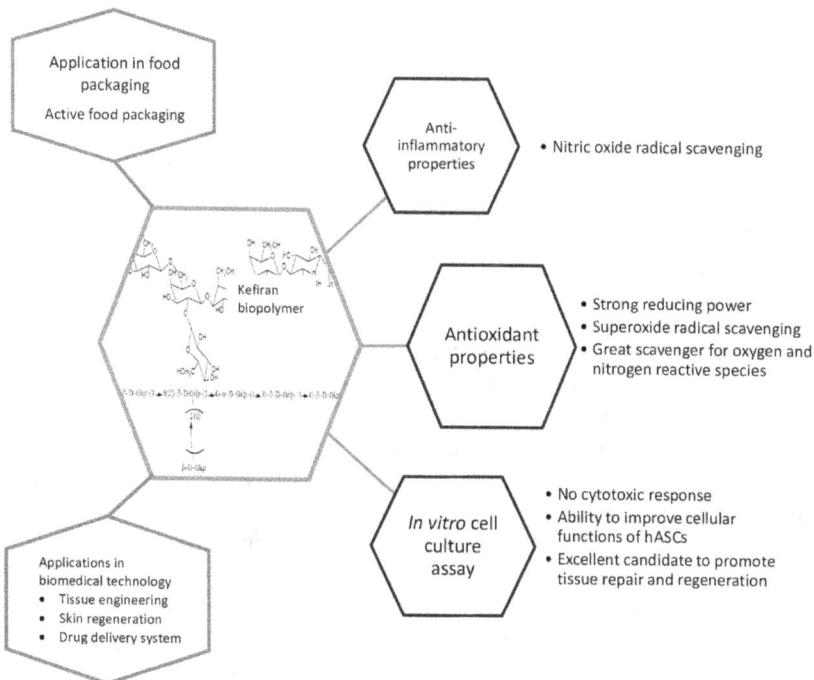

FIGURE 8.2 Different uses of kefiran.

TABLE 8.1
Microbial cellulose applications in various fields.

Food	• Functional packaging
	• Dietary fiber
	• Enzyme immobilization
Pharmaceutical	• Film coating
	• Drug/protein delivery
Biomedical	• Scaffold tissues
	• Vascular grafts
	• Cell therapy
	• Wound care
	• Regenerative tissues
	• Dental implant components
Engineering	• Nanocomposites
	• Agriculture soil conditioner
	• Flat panel display
Environment	• Dye decolorization
	• Heavy metal removal
	• Degradation pollutants
	• Sensors

Source: Based on Cott et al. (2020).

materials carry innovative potential and, as also revealed through the current patent developments in this field, these promising renewable alternatives to conventional petrochemical-based polymers are likely to revolutionize the future of material sciences and material applications.

8.5 PLANTS AS THE SOURCE OF BIOPOLYMERS

Hope for designing, developing, and using eco-friendly biopolymer products—and thus resolving persistent issues relating to microplastic pollution—revolve around plants, the most abundant sources of biopolymers. There are two types of plant-based resources the biopolymers can be derived from for commercial purpose: natural and renewable compounds. The naturally available plant-based biopolymers include plant-based proteins (zein, wheat gluten, soy protein etc.), plant-/algal-based polysaccharides (starch, cellulose, agar, alginate, carrageenan, pectin, konjan, and various gums), lipids (waxes, oils, etc.), polyphenols (lignin, tannin), natural rubber, etc. The renewable compounds as sources of polymer include polylactic acid (PLA), polyhydroxyalkanoates (PHAs), polybutylene succinate (PBS), etc. PHAs include polyhydroxybutyrate (PHB), polyhydroxyvalerate (PHV), and PHB-co-β-hydroxy valerate or poly(3-hydroxybutyrate-co-3-hydroxyvalerate (PHBV). Some of the plant-based polymers, both natural and isolates, together their nature and main uses, are presented in Table 8.2.

TABLE 8.2

Nature and main uses of some plant-based/degradable polymers.

Plant-based natural and renewable polymers and their nature	Main uses
Starch Polymer with glucose monomer units, made of amylase and amylopectin	Production of textiles, paper, synthetic additive usable in medicine, cosmetics, and detergents.
Cellulose Most abundantly occurring natural biopolymer on the planet, made up of glucose monomer units	Highly used in textile industry (e.g., cotton); modern techniques are bringing the polymer into various other commercial applications by altering its physicochemical and mechanical properties.
Polylactic acid (PLA) Derived from sugarcane, corn starch, roots, etc.	High tensile strength and weather- and fracture-resistance, used in building models and as prototype of solid objects and components; raw material in 3D printing.
Polybutylene succinate (PBS) A renewable, biodegradable, and compostable polyester	Food and cosmetic packaging; fabrication of mulching films and for delaying release for fertilizer and pesticides in agriculture; biodegradable drug encapsulation in the biomedical field.
Polyhydroxyalkanoates (PHAs) Natural polyester polymer synthesized by bacteria and archaea	PHA-based biofuels akin to biodiesel except for having high oxygen content and no nitrogen and sulfur (Riaz et al. 2021); potential to create bioplastics with novel properties.
Poly(ℇ-caprolactone) (PCL) A biodegradable polyester with a low melting point around 60°C	Used as an additive for resins to improve their properties, such as impact resistance; applications in medical sciences especially implant techniques; scaffold material for bioartificial vessel prostheses (Wulf et al. 2011).

Quite rich in cellulose, hemicelluloses, and lignin, residues of many plants left after the use of their main product, are ideal sources to derive biopolymers from. These are sugarcane bagasse, rice straw, wheat straw, barley straw, oat straw, corn stalks, sunflower stalks, soybean stalks, cotton stalks, saw dust, etc.

The prevailing naturally occurring reinforcements are based upon the plant species providing jute, flax, hemp, kenaf, etc. With one of the intended combinations of good mechanical properties and low density (1.4 gcm^{-3}), bamboo fibers compete with glass fibers in terms of desirable characteristics such as stiffness and strength at similar volume fractions. El Foujji et al. (2020) have released a study on their screening combinations of various types of bamboo fillers, starting from the raw materials to nanoscale dimensions.

8.6 PRODUCTS BASED ON BIOPOLYMERS CURRENTLY IN MARKET

The unique properties of biopolymers, especially their biocompatibility, biodegradability, and non-toxicity, are alluring for use as alternatives to the conventional plastics. Biodegradable, recyclable, and health-friendly biopolymers are needed in food packaging. The food packaging industry is placing emphasis on the need to develop biopolymer items such as bottles, yogurt cups, candy wrappers, etc. The non-toxic nature of the polymers is also attractive for applications in drug delivery systems in the biomedical field. Biopolymers are also being used for food service ware and food waste bags, coating for papers and cardboards, and fibers for clothing, carpets, sheets, and towels and for wall coverings in houses.

Biopolymers are being probed for their utility in more and more ways, including as nano-sized reinforcements for the enhancement of their properties and for promising practical applications as well.

For each individual application involving a product, environmental safety, social consequences, and cost efficiency need to be examined for a product's entire life cycle in order to adopt the most suited material for the desirable applications with maximum social benefits accrued to the society. Other valuable properties of the product, such as its biodegradability, compostability, etc. must also be tested.

Any product labeled as "biodegradable" cannot be put into use for compost preparation as the same requires the product to reach a thermophilic compost temperature necessary for achieving sanitization (Song et al. 2009) and for undergoing the appropriate degradation process. Some biopolymers, especially used as pots and bags for agronomic or horticultural operations or for waste collection and certified by an authentic agency, are used for home composting if okayed for this purpose or, otherwise, for industrial composting.

Worldwide plastics demand by the packaging sectors is rapidly increasing and expected to reach 600 million tons by 2050 (i.e., 3 times the 2013 volume), according to the final report of the Plastics and Composites Sector (2015). As the biopolymers are environmentally safe and serve as a sustainable alternative to conventional plastics, they are being increasingly used in the packaging industry. The biopolymer-based products are likely to multiply market growth over this period. Europe is ahead of other regions in this regard, accounting for a large chunk of market share in 2020, which is thanks to the rising attendance of key players in the region.

Other products with an ever-increasing demand are natural cosmetics. The personal health-care products made up of natural products rather than those incorporating synthetic plastic products are persuading the cosmetic industry to switch over to safe, bio-based personal care products.

Some of the current facts and figures about the biopolymers market are (Research and Markets 2022) (i) the market value was estimated at $13458.2 million in 2021; (ii) the growth of the market is 13.8% with an estimated value of $33264.7 million by 2028; and (iii) the key companies currently prevailing in the biopolymer market are BASF SE, NatureWorks LLC, Braskem S.A., Novamont S.p.A., Du Pont, Plantic Technologies Ltd., Archer Daniels Midland Company, Danimer Scientific, and Bio-on S.p.A.

The pharmaceutical industry is witnessing a continuously increasing production of biopolymers which are being widely recognized for their wound-healing properties. Natural polymers, like chitosan, pectin, gelatin, and alginate are used in making hydrogels that impart a moist environment for dry wounds. The biopolymers are also utilized for wound dressings.

8.7   FUTURE PROSPECTS AND LIMITATIONS TO OVERCOME

The world is increasingly becoming more and more conscious about the environment. Since human health, well-being, and sustainable progress are all attributes of the environment—directly as well as indirectly—any activity and any process detrimental to the environment is also detrimental to all aspects of the planet's life, with human life being no exception. Conventional plastics are detrimental to the environment in a number of ways. As a matter of fact, conventional plastics are being discontinued, and if they continue to be in use, it is largely thanks to (i) lack of certain biopolymer products, (ii) inadequate supplies or limited availability in the market of biopolymer products, (iii) high prices, or unaffordable costs, of the usable biopolymer items.

So far there are about a dozen big companies engaged worldwide in producing and marketing biopolymer products, new as well as alternatives to non-biodegradable plastics. The biopolymer business appears to be flawed by certain limitations in terms of capacity (inadequate number of branches or management-related problems, for example), supplies of raw materials, poor production efficiency and cost-effectiveness, unacceptable or least preferable product quality, etc. More and more companies—including small-scale ones—need to venture into the biopolymer business. Disposal of the biopolymer products must fetch other ecological and socioeconomic benefits, like through compostability-related properties translated into soil fertility amelioration, leading to enhanced productivity of agro-ecosystems. Research focus should be on searching for new biological resources, new strains of microorganisms, by-products, waste materials/effluents from industrial units, crop residues, etc. to be used for the production of biopolymers. Improvement in technological efficiency of production, product quality, pollution control measures during production processes, and considerable reduction in production costs must be the other objectives of research. Innovations worth every step, of course, must be an inevitable outcome of research intervention. Formulation and implementation of local-to-global level policies for controlling and regulating the biopolymer industry and building up an environment for environmentally safe, ecologically sound, health-friendly, and sustainable systems of polymer production is also essential for the proliferation of the biopolymer business and products.

8.8   SUMMARY

Biopolymers are the polymers derived from natural sources, biosynthesized from living organisms, or chemically synthesized using biological material. Detrimental impacts of conventional plastics underline the necessity for biopolymers. The bio-based polymers are manufactured by using biomass as raw material, that is, by using renewable resources, while petrochemical-based ones are derived from non-renewable and exhaustible resources. The bio-based polymers are mostly biodegradable and compostable and are of multiple uses. Proteins (e.g., zein, gluten, collagen, actin, fibrin, keratin, elastin, silk, etc.) are biopolymers with some 20 amino acids linked with each other through amide bonds. Collagen, the most important structural protein in vertebrates and the most abundant among mammals, is a biopolymer of versatile applications. Silk is a protein-rich biopolymer derived from various silk worm species, for example the mulberry worm *Bombyx mori*. Among the most important and popular bio-based polymers are those made from polysaccharides. Starch is one of the most abundant and interesting raw materials for producing biopolymers. Cellulose, a structural polysaccharide in plants, is almost an inexhaustible polymer usable in the production of eco-friendly and sustainable biopolymers. Chitin is the most abundant aminopolysaccharide polymer and second most abundant polysaccharide in nature after cellulose. Chitosan is the well-known derivative of chitin obtainable through chemical or enzymatic deacetylation of chitin. Polyhydroxyalkanoates are the natural polyester polymers synthesized by bacteria quite akin to conventional petrochemical polymers in their physical properties.

Biopolymer properties like biodegradability and compostability phenomenally help in ameliorating the ecological processes. A biodegradable product is not always compostable, but a compostable one is always biodegradable. Biodegradability indicates that a material will fully degrade within a specified period of its disposal. A compostable type of polymer is supposed to be biodegradable into "soil conditioning material". Microbial biomass originating from fungi, especially yeast and mycelium biomass, has enormous potential for use in various technologies. Kefir grains and bacterial cellulose are of wide industrial applications for producing a variety of biodegradable biopolymers usable for various purposes. Microbial cellulose, especially bacterial cellulose, has multiple properties of a biopolymer with several applications in so many fields, notably foods, pharmaceuticals, biomedical sciences, engineering, and the environment. The naturally available plant-based biopolymers include plant-based proteins (zein, wheat gluten, soy protein, etc.), plants/algal-based polysaccharides (starch, cellulose, agar, alginate, carrageenan, pectin, konjan, and various gums), lipids (waxes, oils, etc.), polyphenols (lignin, tannin), natural rubber, etc. The renewable compounds as sources of polymers include polylactic acid (PLA), polyhydroxyalkanoates (PHAs), polybutylene succinate (PBS), etc. PHAs include polyhydroxybutyrate (PHB), polyhydroxyvalerate (PHV), and PHB-co-β-hydroxy valerate (PHBV). The prevailing naturally occurring reinforcements are based upon the plant species providing jute, flax, hemp, kenaf, etc. The non-toxic nature of the polymers has made their applications attractive in food packaging, the pharmaceutical and cosmetic industries, and in the biomedical field. Biopolymers are also being used for food service ware and food waste bags, coating for papers and cardboards, and fibers for clothing, carpets, sheets, and towels and for wall coverings in houses and are being probed for their utility in more and more ways. Natural polymers, like chitosan, pectin, gelatin, and alginate are used in making hydrogels. The conventional plastics are being increasingly discarded. The biopolymers business appears to be flawed by certain limitations. More and more companies—including small-scale ones—need to venture into the biopolymer business. Disposal of the biopolymer products must fetch multiple ecological and socioeconomic benefits. Research focus should be on searching for new biological resources, new strains of microorganisms, by-products, waste materials/effluents from industrial units, crop residues, etc. to be used for the production of biopolymers. Research should also focus on improvement in technological efficiency of production, product quality, pollution control measures during the production processes and considerable reduction in production costs.

REFERENCES

Angelina, A. and Vijayendra, S. V. N. 2015. Microbial polymers: The exopolysaccharides. In: Kalia, V. (ed.) *Microbial Factories*. New Delhi: Springer. doi: 10.1007/978-81-322-2595-9_8.

Azuma, K, Ifuku, S., Osaki, T., Okamoto, Y. and Minami, S. 2014. Preparation and biomedical applications of chitin and chitosan nanofibers. *Journal of Biomedical Nanotechnology*, 10(10):2891–2920. doi: 10.1166/jbn.2014.1882.

Cerimi, K., Akkaya, K. C., Pohl, C., Schmidt, B. and Neubauer, P. 2019. Fungi as source for new biobased materials: A patent review. *Fungal Biology and Biotechnology*, 6:17. doi: 10.1186/s40694-019-0080-y.

Cottet, C., Ramirez-Tapias, Y. A., Delgado, J. F., de la Osa, O., Salvay, A. G. and Peltzer, M. A. 2020. Biobased materials from microbial biomass and its derivatives. *Materials*, 13:1263. doi: 10.3390/ma13061263.

El Foujji, L., El Bourakadi, K., Qaiss, A. and Bouhfod, R. 2020. Characterization and properties of biopolymer reinforced bamboo composites. In: Jawaid, M., Mavinkere Rangappa, S. and Siengchin, S. (eds) *Bamboo Fiber Composites: Composites Science and Technology*, 147–173. doi: 10.1007/978-981-15-8489-3_9.

Elieh-Ali-Komi, D. and Hamblin, M. R. 2016. Chitin and chitosan: Production and application of versatile biomedical nanomaterials. *International Journal of Advanced Research* (Indore), 4(3):411–427. www.ncbi.nlm.nih.gov/pmc/articles/PMC5094803/

Encalada, K., Aldás, M. B., Proaño, E. and Valle, V. 2018. An overview of starch-based biopolymers and their biodegradability. *Revista Ciencia e Ingeniería*, 39(3):245–258. www.researchgate.net/publication/328773263_An_overview_of_starch-based_biopolymers_and_their_biodegradability

Guimaraes, L. H. S., Peixoto-Nogueira, S. C., Michelin, M., Rizzatti, A. C. S., Sandrim, V. C., Zanoelo, F. F. and Polizeli, M. D. L. 2015. Screening of filamentous fungi for production of enzymes of biotechnological interest. *Brazilian Journal of Microbiology*, 37:474–480. doi: 10.1590/S1517-83822006000400014.

Gupta, P and Nayak, K. K. 2014. Characteristics of protein-based biopolymer and its application. *Polymer Engineering and Science*, 55:485–498. doi: 10.1002/pen.23928.

Kadimaliev, D., Kezina, E., Telyatnik, V., Revin, V., Parchaykina, O. and Syusin, I. 2015. Residual brewer's yeast biomass and bacterial cellulose as an alternative to toxic phenol-formaldehyde binders in production of pressed materials from waste wood. *Bioresources*, 10:1644–1656.

Liu, Y., Ahmed, S., Sameen, D. E., Wang, Y., Lu, R., Dai, J., Li, S. and Qin, W. 2021. A review of cellulose and its derivatives in biopolymer-based for food-packaging application. *Trends in Food Science and Technology*, 112:532–546. doi: 10.1016/j.tifs.2021.04.016.

Mokhtari, S., Jafari, S. M., Khomeiri, M., Maghsoudlou, Y. and Ghorbani, M. 2017. The cell wall compound of Saccharomyces cerevisiae as a novel wall material for encapsulation of probiotics. *Food Research International*, 96:19–26. doi: 10.1016/j.foodres.2017.03.014.

Peltzer, M. A., Salvay, A. G. and Delgado, J. F., de la Osa, O. and Wagner, J. R. 2018. Use of residual yeast cell wall for new biobased materials production: Effect of plasticization on film properties. *Food and Bioprocess Technology*, 11:1995–2007. doi: 10.1007/s11947-018-2156-8.

Plastics and Composites Sector. 2015. *Smart Technologies for Smart Industries Targeted for Smart Applications for Smart Communities*. Academy of Sciences Malaysia. 133pp. https://bm.akademisains.gov.my/download/ms3.0/Plastics_Composites_Industry_Sector.pdf

Radhouani, H., Bicho, D., Gonçalves, C, Maia, F. R., Reis, R. L. and Oliviera, J. M. 2019. Kefiran cryogels as potential scaffolds for drug delivery and tissue engineering applications. *Materials Today Communications*, 20:100554. doi: 10.1016/j.mtcomm.2019.100554.

Radhouani, H., Gonçalves, C, Maia, F. R., Oliveira, J. M. and Reis, R. L. 2018. Biological performance of a promising kefiran-biopolymer with potential in regenerative medicine applications: A comparative study with hyaluronic acid. *Journal of Materials Science: Materials in Electronics*, 29:124. doi: 10.1007/s10856-018-6132-7.

Research and Markets. 2022. *Biopolymers Market, by Product Type, by Application, by End User, and by Region—Size, Share, Outlook, and Opportunity Analysis, 2021–2028*. New York: Insights Pvt Ltd. 170pp.

Riaz, S, Rhee, K. Y. and Park, S. J. 2021. Polyhydroxyalknoates (PHAs): Biopolymers for biofuel and biorefineries. *Polymers (Basel)*, 13(2):253. doi: 10.3390/polym13020253.

Sabu, C., Mufeedha, P. and Pramod, K. 2019. Yeast-inspired drug delivery: Biotechnology meets bioengineering and synthetic biology. *Expert Opinion on Drug Delivery*, 16:27–41. doi: 10.1080/17425247.20119.1551874.

Singh, V. 2019. *Fertilizing the Universe: A New Chapter of Unfolding Evolution*. London: Cambridge Scholars Publishing. 285pp.

Singh, V. 2020. *Environmental Plant Physiology: Botanical Strategies for a Climate Smart Planet*. Boca Raton: CRC Press (Taylor and Francis). 216pp.

Song, J. H., Murphy, R. J., Narayan, R. and Davies, G. B. H. 2009. Biodegradable and compostable alternatives to conventional plastics. *Philosophical Transactions of the Royal Society B - Journals*, 364(1526):2127–2139. doi: 10.1098/rstb.2008.0289.

Šprajcar, M., Horvat, P. and Kržan, A. 2012. *Biopolymers and Bioplastics: Plastics Aligned with Nature*. European Union. 32pp. www.umsicht.fraunhofer.de/content/dam/umsicht/de/dokumente/ueber-uns/nationale-infostelle-nachhaltige-kunststoffe/biopolymers-bio-plastics-brochure-for-teachers.pdf

Tang, X. and Alavi, S. 2011. Recent advances in starch, polyvinyl alcohol based polymer blends, nanocomposites and their biodegradability. *Carbohydrate Polymers*, 85(1):7–16. doi: 10.1016/j.carbpol.2011.01.030.

Vijayendra, S. V. N. and Shamala, T. R. 2014. Film forming microbial biopolymers for commercial applications: A Review. *Critical Reviews in Biotechnology*, 34:338–357. doi: 10.3109/07388551.2013.798254.

Wulf, K., Teske, M., Löbler, M., Luderer, F., Schmitz, K.-P. and Sternberg, K. 2011. Surface functionalization of poly(ε-caprolactone) improves its biocompatibility as scaffold material for bioartificial vessel prostheses. *Journal of Biomedical Materials Research*, 98(1):89–100. doi: 10.1002/jbm.b.31836.

9 Global Legislature for the Mitigation of Microplastics in the Marine Environment

The gravity of the microplastics pollution in the marine environment calls for chalking out concrete and workable strategies and their effective implementation at global, regional, and national scales. Effective and workable strategies are, of course, identified and a global legislature framework has also been developed. The Earth's marine ecosystems, however, are unabatedly reeling under crises of microplastics pollution. The issues pertaining to this formidable situation are still to be resolved.

Open marine ecosystems are independent solar-powered ecosystems and coastal ecosystems are among the most productive ones. Changes in marine and coastal ecosystems induce changes in terrestrial ecosystems due to interconnectedness among ecosystems. Spoilage of these aquatic ecosystems owing to microplastic pollution affects marine, terrestrial, and human lives in a number of ways. The marine ecosystems impart many valuable services to a variety of life—both aquatic and terrestrial— as well as to human societies. These include provisioning, regulating, supporting, and cultural services (Figure 9.1). Therefore, constitution and implementation of global legislatures for microplastic mitigation has become of critical significance.

9.1 INTERNATIONAL REGULATORY BODIES AND REGULATIONS

Plastics, when in use and in being reduced to wastes, are lost to all environments across their entire value chain (Boucher and Friot 2017; UNEP 2018). Mismanagement of municipal solid waste (MSW), especially in economically poor countries, is the major cause of the macroplastic loss into all environments. This fate of the plastics raises challenges as well as opportunities to prevent leakage into natural and technical systems (Ryberg et al. 2019).

An unabated increase of microplastics pollution is an eclipse on the very sustainability of the living planet. Multiplying at an alarming rate, this vicious problem unfolded itself much later than other environmental problems and is an undeniable ethical issue to be resolved with effective and workable measures. Many research workers, environmentalists, ecologists, social activists, common people, and institutes throughout the world have been pressuring political leaders and governments to resort to action plans to control the situation advancing towards ecological disaster. On growing public demand, governments and organizations are all making efforts to address macro- and microplastic-related issues by applying sustainability principles and regulations

DOI: 10.1201/9781003312086-9

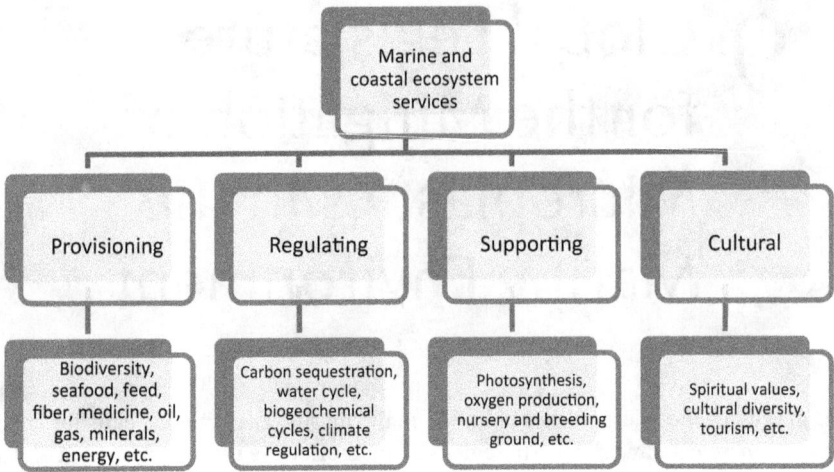

FIGURE 9.1 Various marine ecosystem functions translating into ecosystem services vital for the planet's life.

(Mitrano and Wohlleben 2020). It can be easily grasped in mind that increasing microplastics' presence in marine and other aquatic ecosystems, soils, and eventually into public life encumbers sustainability. One of the stark examples is incorporating plastics in a circular economy as the European Union attempts to do. Embracing plastics as one of the demonstrative examples, the linear economy, as opposed to the circular economy, has been and continues to be laden with numerous shortcomings, including serving as a precursor of environmental disruptions. Plastics, and specially the microplastics, as Backhaus and Wagner (2019) emphasize upon, should be in the focus of communication pertinent to the issue of environmental sustainability.

There are many ways for governance to resort to reduction in plastic use, consumption, and disposal practices. However, once macroplastics in the plastic wastes or during their various uses are fragmented into micro- and nanoplastics and their entry into the environment succeeds, their control through governance measures is difficult to realize. Macroplastics management, as an essential part of a circular economy, as Mitrano and Wohlleben (2020) suggest, would help resolve a major part of the issue relating to microplastics pollution caused by secondary microplastics, that is, the microplastics originating as a result of fragmentation from the plastic wastes in waters and soils. Primary microplastics, such as microbeads manufactured through engineering means, are considered a relatively less severe problem compared to that emerging from secondary microplastics. The proposed microplastics regulations, however, especially revolve round the primary microplastics.

Sustainable Development Goal 14 (SDG 14), one of the 17 SDGs propounded by the United Nations General Assembly in 2015, focuses on the conservation and sustainable use of oceans, seas, and marine resources for sustainable development (United Nations 2017). Out of the ten targets defined by the United Nations for SDG 14, the first one, "reducing marine pollution", includes reducing impacts from plastic pollution as covered under indicator 14.1.1b, although at least 12 SDGs are directly or indirectly affected by the pollution caused by microplastics (Walker 2021). Before then, several strategies to prevent and reduce microplastics pollution and regulate plastic production and use had come into existence. The UN Convention on the Law of the Sea (UNCLOS) considered in 1982 contains two articles—Article 207 and 211—that focus on the concern relating to the decrease, control, and prevention of plastic waste in marine ecosystems. Again, the United Nations emphasized and called for the world's states to prevent and significantly reduce marine pollution of all sorts, especially that caused due to land-based activities, including floating plastic debris.

The main declarations of the UN General Assembly in connection with marine environment include constitution of public-private sector partnerships towards addressing ecological and socioeconomic impacts of plastic pollution by aligning with national strategic frameworks. The UN resolutions seek comprehensive and integrated efforts of international, regional, and national organizations towards finding solutions for marine pollution arising due to increasing plastic litter accumulation. The prominent international organizations include United Nations Environment Program (UNEP), Food and Agriculture Organization (FAO), and International Maritime Organization (IMO). These are assisted by sub-regional fisheries management organizations.

Plastics pollution, as also claimed by UNEP (2011), is a critical environmental issue. The UN Convention on Sustainable Development (Rio +20) highlighted the urgency of plastics pollution control in the oceans. In the strategies to resolve plastics pollution related issues, the IMO also came into the limelight. Management of marine pollution was the major objective to be achieved by 2025. International Convention for the Prevention of Marine Pollution (MARPOL) is the principal legislature's body to look into various objectives set out for marine pollution management, including plastic pollution control and prevention.

One of the crucial plastic/microplastic control measures that stems from Article 70 of the Convention on Biological Diversity (CBD) is reduction of plastic pollution impacts on marine biodiversity by means of the strategies, including the environmental impact assessment (EIA) and strategic environmental assessment (SEA). Key decisions aimed at controlling plastic debris in marine and coastal environments were declared at the 16th meeting of the Subsidiary Body on Scientific, Technical and Technological Advice (SBSTTA), the Scientific Advisory Body to CBD, as follows (Thushari and Senevirathna 2020):

1. Monitoring and documentation on plastic waste impacts on ecosystem and biological diversity.
2. Scientific research on plastic debris control and management.
3. Capacity building programs relating to methods of prevention and control of plastic accumulation.

The Convention on the Conservation of Migratory Species (CMS), or the Bonn Convention—a United Nations Environmental Treaty—is assigned many tasks relating to plastic pollution control and prevention. These are searching marine debris hotspots, assessment of plastic waste impacts on marine biodiversity, identifying ways and means of controlling waste accumulating sources at the regional level, and national level implementation of an action plan to manage waste accumulation in marine ecosystems.

UNEP (2011) emphasizes on an integrated waste management approach that also includes the management of plastics waste in the marine environment. This approach seeks to resolve serious issues pertaining to microplastic pollution pervading most of the planet. One of the prominent pieces of international legislation stems from the Basel Convention on the Control of Transboundary Movement of Hazardous Wastes and Their Disposal that was endorsed in 1989 and brought into effect in 1992. This convention put into action the Bali Declaration on Waste Management for Human Health and Livelihoods in 2008 (Thushari and Senevirathna 2020). As the macro- and microplastics are parts of the hazardous wastes in the marine environment, their mitigation is also covered in the Basel Convention followed up by the Bali Declaration. UNEP's Global Partnership on Waste Management (GPWM) has paved ways to waste management in collaboration with international organizations and non-government organizations (NGOs). Yet another attempt at significant plastic waste management with the collaboration of UNEP and NOAA (National Oceanic and Atmospheric Administration) lies in the Honolulu Strategy—a global-level strategic framework aiming at prevention and management of marine debris.

9.2   REGIONAL REGULATORY BODIES AND REGULATIONS

UNEP's Regional Sea Program has forwarded a proposal about an action plan for 13 regional seas (Thushari and Senevirathna 2020): Eastern African Sea, East Asian Sea, Gulf of Aden, Red Sea, Caspian Sea, Northeast Atlantic Sea, Mediterranean Sea, Baltic Sea, Northwest Pacific Sea, Southeast Asian Sea, Wider Caribbean Sea, Black Sea, and South Asian Sea. Coastal cleanup programs as global projects in the selected regional seas have been accomplished.

EU's Marine Strategy Framework Directive (MSFD), which came into being in 2008 as an effort of collaboration among all the nations in the European Union, is aimed at reducing marine wastes to the minimum at a regional level. NOAA is looking after the Hawaii and South African coasts for waste management. Estuarine ecosystems in South America are being looked into for the effective management of plastic pollutants.

In the ambit of new REACH (European Union law for Registration, Evaluation, Authorization and Restriction of Chemicals) regulations dating from December 18, 2006, microplastic pollution is currently an issue of chemical risk management. Earlier, in a bid of safe chemical use, the REACH regulations embraced the "no data—no market" principle. Considering that owing to their high molar mass polymers' bioavailability was not a case, the polymers were not included in the registration regime in the original REACH regulations. Plastic regulations now may individually include polymers and additives (Mitrano and Wohlleben 2020).

Additives as small molecules used in polymers, as per REACH registration, are needed to undergo tests against their deleterious effects on human health as well as on the environment. The Environmental Protection Agency of USA (USEPA) applies the concepts of polymers of low concern (PLC), that is, the polymers assumed to impart non-significant effects on human health and environment. Most of the jurisdictions outside the European Union are also applying the same concept as USEPA is doing (ECETOC 2019). Solid plastics and other polymers, according to Mitrano and Wohlleben (2020), are considered PLCs provided they: (i) fulfill specific criterion relating to molar mass and low-molar-weight compounds (<1000 and gmol^{-1}); (ii) do not fulfill exclusion criterion, for example, pre-defined reactive functional groups or chemical elements, biodegradation, and/or water solubility, and cationicity; and (iii) are produced in high tonnage, except PVC.

The existing definition of primary microplastics is a constraint to their regulations. The European Chemical Agency (ECHA), therefore, redefines the microplastics as follows: a material comprising solid polymer containing additives and other substances, in which $\geq$1% w/w of particles have (i) all dimensions 1 nm $\leq$ x $\leq$ 5 mm, or (ii) for fibers, a length of 3 nm $\leq$ x $\leq$ 15 mm and length-diameter ratio of >3.

This broad "regulatory" definition (Figure 9.2) helps create a scope for microplastic regulation somewhat more strictly. An example of contemporary plastic regulations for various polymer types, additives used, and solid primary microplastics at the regional level (European Union) are presented in Figure 9.3.

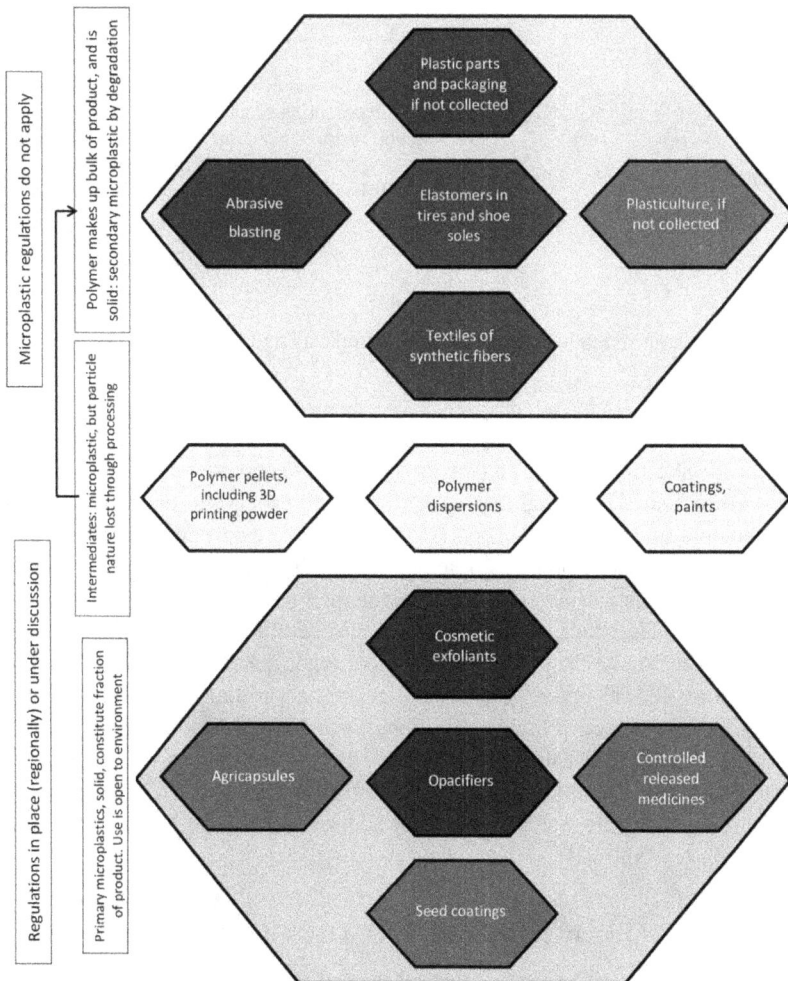

FIGURE 9.2   Categorized boundaries of microplastic regulation.

Source: Based on Mitrano and Wohlleben (2020).

Regulation of polymers, including functional polymers and solid plastics	• Exempted from REACH due to low bioavailability
Regulation of additives	• REACH, if > 1.0 ton manufactured or imported
Regulation of primary solid microplastics	• Proposed REACH restriction, potentially from 2022: banned if dispersed in environment; labeling and reporting if used only industrially and/or losing particle nature in application; biodegradable, or natural, or soluble polymers exempted

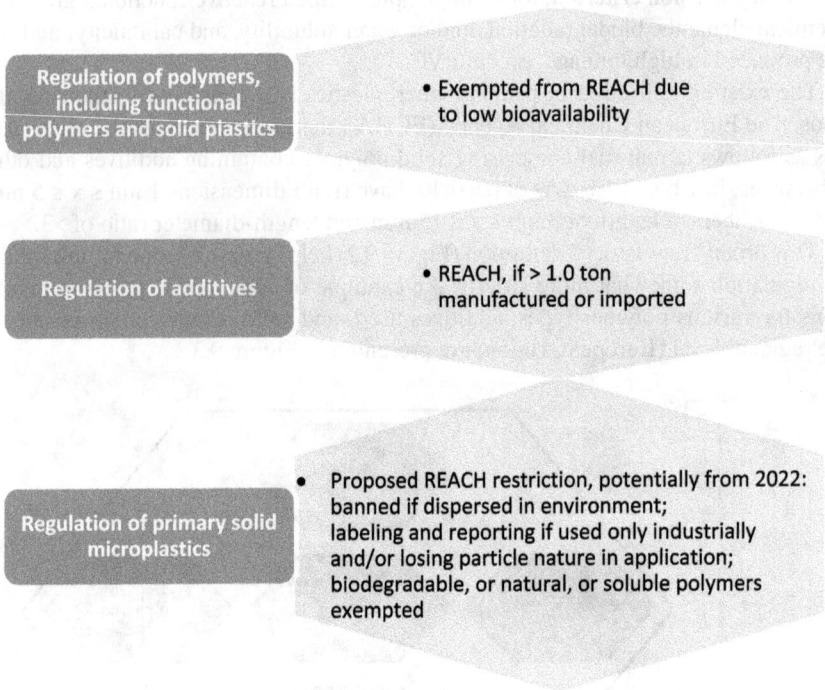

FIGURE 9.3   Current plastic regulation in the European Union.

Source: Based on Mitrano and Wohlleben (2020).

All the plastic products falling under the ECHA's broad microplastic definition would need to bear a label and their uses need to be brought to the notice of ECHA. Natural, soluble, and biodegradable polymers, however, are outside the scope of any regulation. The intermediate products, like pellets and dispersions, fall in the category of such restrictions. The plastic products not losing their particle nature during processing are to be banned. The beads used in cosmetics, agricultural capsules, and seed coatings, thus, are counted among the products under the category of those marked to be banned.

As much as 42,400 tons of the plastics released into the environment in a year are likely to be reduced following this approach (Mitrano and Wohlleben 2020). Reduction in this much plastic waste into the environment, however, does not appear to be substantial, as the microplastic releases from synthetic textiles, tire wear, and waste fragmentation, etc. are not within the scope of ECHA's comprehensive microplastic regulatory definition.

9.3   REGULATORY BODIES AND REGULATIONS IN INDIA

India is among the world's largest plastic consumers and, therefore, one of the largest plastic waste producers and is most affected by the environmental pollution caused by macro- and microplastics, especially in coastal ecosystems. According to the Central Pollution Control Board (CPCB) of India, the country produces as much as 26,000 ton plastic waste per day. The first major regulatory directions by India emanate from the Environment (Protection) Act, 1986. The main purpose of the Act was to provide protection to the environment, including management of all sorts of wastes.

India's first rule regarding plastic wastes was the Recycled Plastics (Manufacture and Usage) Rules 1999. Controlling food packaging using recycled plastics and managing litter problems was the major aim. The main dictates of the rule were:

1. All carry bags of 20 micron size were banned.
2. Recycled and virgin-colored polythene bags would be used only for non-food applications, not for the packaging of food products.
3. The guidelines for plastic recycling were made mandatory.

The Bureau of Indian Standards (BIS) is the body that issues necessary regulations pertaining to plastic, polyethylene terephthalate (PET), and PET recycle usage. The document IS 14534:1998, *Guidelines for Recycling of Plastics*, provides detailed processes (viz., selection, segregation, and processing of waste/scrap) for the recycling and recovery of the plastic wastes. In addition, BIS provides the lists suggesting end products of the recycled plastics for use for different purposes. In 1999, the BIS brought into effect the rules imposing prohibition on the usage of recycled plastics for producing bags, sacks, and wrapping material largely used for food packaging (Plastics for Change 2020).

The Plastics (Manufacture, Usage and Waste Management) Rules 2009 were passed subsequent to those of 1999 to regulate plastic wastes more effectively. These Rules regulate the plastics in India in the following ways:

- Carry bags and containers designed from virgin plastics with no coloring agent.
- Prohibition of the carry bags/containers manufactured using recycled or biodegradable plastics for food packaging.
- Use of pigments/colorants in carry bags/containers made up of recycled/ biodegradable plastics usable for the reasons other than food packaging as per the BSI specifications.
- Prohibition of manufacturing, storage, distribution or sale of carry bags of virgin/recycled/biodegradable plastics <30 × 45 cm in size and <40 microns in thickness.
- No manufacturing of carry bags/containers/pouches or multi-layer packing using biodegradable plastic materials that mismatch with BIS standards under "Specifications for Compostable Plastics".
- Ban on manufacturing, stocking, distribution, or sale of non-recyclable laminated plastics, etc. other than food packaging.
- Plastic recycling to be carried out as per BIS Specifications: IS 1453: 1988—*The Guidelines for Recycling of Plastics*.

Plastic Waste (Management and Handling) Rules 2011 was the next ruling the Indian government promulgated. It has the following main elements:

- Minimum plastic bags thickness increased to 40 microns.
- Recycled carry bags manufactured using compostable plastics as per BIS standards.
- Ban on plastic sachets for storage, packaging, or sale of tobacco-based products.
- Waste pickers, agencies, or groups to be employed by municipalities working for waste management.

The Government of India implemented the Plastic Waste Management Rules in 2016 which were further amended in 2018. Involving the Urban Local Bodies (ULBs), institutions, and residential, commercial, and defense establishments, the regulatory network provides guidelines for effective disposal and management of the plastic wastes.

The Government of India updated the rules with the Plastic Waste Management Amendment Rules, 2021, imposing bans on manufacturing, importing, maintaining stock, distribution, sale, and usage of identified single-use plastic (SUP) items in effect from July 1, 2022. The plastic items to be prohibited include polystyrene, expanded polystyrene, ear-buds with plastic sticks, cutlery, plates, cups, candy sticks, etc.

The major responsibilities to implement the rules have been assigned to CPCB and all the State Pollution Control Boards (SPCBs).

9.4   IMPACT OF THE CONSUMER VOICE AND BEHAVIOR ON POLICY DEVELOPMENTS

Peoples of the world are beginning to feel the pressure to ban selected plastic products like carry bags and disposable cups. As the plastic pollution is increasingly becoming uncontrollable and new policies for putting a brake on this problem are coming into light, people are becoming increasingly conscious about it. Frequent usage and disposal of single-use plastic is, undoubtedly, the main issue of public interest. Examples of overwhelmingly used single-use plastics that contribute to spoiling human habitats and affecting public health are plastic bottles (including plastic lids), food wrappers, cigarette butts, sanitary items, stirrers, cutlery, etc. Presence of these consumer items in fresh and marine waters constantly fragmenting into microplastics is advancing towards the climax of an ecological mess.

Marazzi et al. (2020), based on their study on as many as 27 plastic reduction actions on some consumer plastic products of common use that end up in the freshwater ecosystems in Europe, observed that there is enormous potential of reducing plastic/microplastic loads in the environment (Table 9.1). What we can beautifully elicit from this study is that the consumers-citizens' role in reducing plastic pollution is phenomenal. The policies and legislations brought into effect, in fact, are a bottom-up process. These are basically suggested by the affected populations directly or through impacts on their environments, resources, and livelihoods. Positive changes in the behavior of the consumers are induced on how the policies stimulate and

TABLE 9.1

Potential effectiveness of consumers' actions to reduce pollution caused by most common plastic items in European freshwater environments.

Plastic items dealt by consumer actions*	Potential impact (tons of plastics saved per year)
1. Plastic bottles (including plastic lids)	6,741 + Unknown
2. Food wrappers	Unknown
3. Cigarette butts	2,482
4. Food takeaway containers	1,290
5. Cotton bud sticks	61
6. Plastic cups	4,500
7. Sanitary products	43,749 + Unknown
8. Packaging material for the smoking products	Unknown
9. Plastic straws, stirrers, cutlery, etc.	5,510.2
10. Plastic bags	9,000

Source: Based on Marazzi et al. (2020).

Note: *Consumer actions: 1. Reusable water bottles; direct delivery of milk to households; refilling of bottles/ containers of detergents, shampoos, drinks, etc.; and drinks in cardboard containers; 2. Correct disposal; 3. Correct disposal; 4. Reusable takeaway containers; 5. Use of paper in place of plastic sticks in cotton buds; 6. Reusable plastic cups, reusable glass cups, and reusable bamboo cups; 7. Not flushing wet wipes, menstrual cups, biodegradable wet wipes, etc.; 8. Correct disposal; 9. Wooden cutlery, reusable cutlery, wooden stirrers, paper straws, reusable straws (bamboos/glass/steel/silicon); 10. Reusable cotton tote bags.

benefit people in one way or the other. Seeking, ensuring, and enhancing consumers' participation for rectifying a serious problem arising out of consumer products must be central to the policies on plastic regulations.

9.5 IMPACT OF THE INDUSTRIAL PRACTICES ON POLICY FRAMEWORK

Most countries have laws to control pollution created by industries. The firms engaged in production processes for products distribution in market systems to pass on to consumers require possessing environmental permits. The permits include categorical provisions about waste management, bans on hazardous plastic products/materials, and incentives for effectively adopting recycling practices. Increasing recyclability of the used plastic materials by the use of a few polymers and chemical additives are the provisions set for the industries/firms. With a positive trade balance for plastics that gives an impression of plastic manufacturing at a local level being substantial, and with the industries largely consisting of smaller or informal enterprises, what is required is to ascertain minimum plastic loss during plastic products processing and transport, which, in turn, requires formulation and implementation of policies. Unlike in the developed countries, national legal frameworks in the developing countries normally have no regulation on pellet spills from the industries. The European nations, for instance, have formulated legislative measures for the plastic regulation on production, transportation, and pellets use by means of EU Packaging Directive (Directive 2008/98/EC, 2008), REACH (Regulation (EC) No 1907/2006), and the Industrial Emissions Directive (Directive 2008/98/EC) (Alpizar et al. 2020). In the EU, all plastic items would be recycled by 2030 (Gilli et al. 2018), a policy which is largely to be influenced by industrial behaviors.

Plastic particles emissions, as the whole world witnesses, go unabated. Checking the emissions is solely part of the industry's responsibility. However, the industry sector did not translate its responsibility into action. Ethics is put behind the profitability in a capitalist system. Checking or minimizing emissions makes imposition of legislative measures necessary. Taxing the plastic producing industrial units on the basis of per unit emissions ending into marine ecosystems is supposed to be an ideal system. However, such emissions from an individual firm are difficult to be measured. Therefore, the taxes should be imposed on the basis of indirect environmental deterioration caused by the plastic items and plastic production processes or on how recyclable the plastic products' components are (Alpizar et al. 2020). Plastic pollution originating from products like PS and PVC are more harmful to health and so deserve higher tax rates. If an industrial unit is using certain chemical additives added to plastic products that are hazardous to public health, it needs to be taxed. In sharp contrast are the subsidies to be provided to the plastic industry if they strictly follow the rules and regulations and meet specific criteria set by governing bodies and significantly contribute to reducing plastic/microplastic pollution.

A right-based action used for targeting the plastic industry is what is known as Extended Producer Responsibility (EPR). The EPR concept attempts to address the responsibilities or the maintenance of a greener and cleaner environment even if the production chain is completed. The firms manufacturing plastic items and packaging materials are given encouragement for collecting used packing materials and recycling by means of funding and operational activities.

It is mainly being implemented in the European Union. Other developed countries, notably Canada and Japan, also follow this approach. The developing countries so far are not very serious about this development. Under EPR, the property rights and duties are assigned to the plastic product producers who have to treat and dispose of the products after consumption. The EPR is considered to be one of the best practices paving the way to minimum plastic waste accumulation rate in the environment. Encouraging plastic waste volume reduction in the usage of virgin material and developing a recycle sector are the major goals intended by adopting EPR (Brouillat and Oltra 2012). The Sustainable Material Management (SMM) is the other important plan for giving an appropriate response to pollution, including plastic pollution control, meant for a clean environment (Thushari and Senevirathna 2020). Developed countries, such as Japan where this program has been going on since 1997, are especially fond of this program.

When the plastic industry's behavior comes into consideration, the proviso of information to an industrial unit as well as to customers should strictly be there. If a firm or an industrial unit works in tune with environmental safety norms, consumers and regulatory bodies would welcome it. Studies focused on corporate social responsibility suggest firms are interested in environmental issues (Alpizar et al. 2020). If the industrial behavior is antagonistic to environment and consequently to public health, consumers are driven to change their behavior. They would raise their voice to garner support to impose pressures on the industry to change behavior in favor of the environment and public health. One interesting fact we can elicit from the industry-consumer relationship is that the consumers are conscious about the environmental performance of the industry. A deposit-refund system for plastic items like bottles and containers will be phenomenal towards reducing plastic waste loads in the environment.

With the burgeoning microplastic pollution of marine ecosystems, it is an obligation for the plastic industry units to switch over to producing environment-friendly substitutes of conventional plastic products. Such alternatives include barrier coatings, biodegradable plastics, compostable plastics, bioplastics, and biodegradable polymers for food packaging.

9.6 NEED TO PRIORITIZE AND CHANGE REGULATIONS BASED ON INNOVATIONS AND IMPACT ON MARINE ENVIRONMENT

Of the several thousand million metric tons of the plastics manufactured so far in the world, about 80% has ended up in marine environments and landfills. Several ways and legislative measures of mitigating plastics/microplastics pollution have so far been tried. The problem, however, goes on persisting and magnifying. One of the approaches to handle the problem more effectively is to set priorities and change plastic regulations based on innovations and relative impact on the marine environment.

Polyurethane (PUR), the world's third most widely used plastic, is of multifaceted attributes we overwhelmingly depend on in our day-to-day life. This plastic demands energy-intensive production processes based on petrochemicals. The petrochemical feedlots make up the polyols PUR structures are made up of. The polyols, in turn, are made up of epoxides through the process like oxidation or hydrochlorination simultaneously responsible for a large carbon footprint. If an alternative feedstock is used in the process, it would lead to an innovative approach for the production of PUR products.

Use of atmospheric CO_2 as a feedstock in the process of PUR plastic production possible due to advances in catalytic science would be a revolutionary approach. Three tons of atmospheric CO_2 are utilized (removed from the atmosphere) for each ton of epoxide replaced with CO_2 as a raw material. Using CO_2 in the PUR production process is promising on three fronts: (i) plastic production with significant reduction in environmental impact of the plastic materials, (ii) contribution to the removal of atmospheric carbon, the excess of which is responsible for global warming and climate change, and (iii) production of the items of socioeconomic significance.

In fact, polyols making use of CO_2—known as polyethercarbonates—are increasingly becoming common in many companies willing to reduce their carbon footprints. The extra advantages of using CO_2 as a raw material during the PUR production process is significant improvement in product quality: improved flame retardance properties, and enhancement in the chemical, temperature, and hydrolytic resistance in the coating, adhesive, sealant, and elastomer properties (Blackburn

2018). Further, CO_2-dependent processes of plastic production is comparatively more cost-effective than that of the conventional petrochemical-based one.

The plastic regulation bodies, therefore, must provide incentives to the firms that switch over to the alternative environment-friendly processes of plastic production. Such innovative processes must be given priority in the legislations relating to plastic production.

Applications of nano-engineering for the purpose of creating recyclable materials enabling replacement of complex non-recyclable, multi-layered packaging may be brought into use. This practice has been applied successfully by a team at the University of Pittsburgh. Such promising packaging materials mimic the natural ones using only a few molecular building blocks for creating amazing varieties of material (Iles 2018).

A magnetic additive can create air and moisture insulation effectively and thus save food items—especially some sensitive ones including coffee and medicines—from getting spoiled.

Compostable and multi-layered materials made from agriculture and forest by-products could be pretty feasible for the packaging of food products, especially fresh fruits and nuts as well as for food grains. The agriculture and forestry by-products, the products of photosynthesis, are rich in cellulose, the most abundant polymer on the planet, which is a biodegradable as well as environment-friendly alternative of the conventional petroleum-based plastic packaging material.

As far as the recycling of plastics is concerned, only 14% packaging materials are generally collected and just 2% appropriately recycled (Iles 2018). Recycling of plastic wastes should have a priority in the plastic regulation laws before the wastes end up in oceans and seas and become a source of microplastic pollution.

Innovations in plastic wastes management and microplastic pollution mitigation need to be incorporated in the principles of the circular economy. The *New Plastics Economy* report released by the Ellen MacArthur Foundation in 2016 had warned that by 2050 the oceans might contain more plastic wastes than fish, by weight (Iles 2018). The warning needs to be taken seriously and plastics management approaches must be implemented sincerely and effectively. What is urgently needed is the permanent solution of the root cause of the problem. Redesigning the plastic manufacturing, marketing, and utilization systems in tune with the concepts of a circular economy would help phenomenally change mindsets but would also require a new mindset. The essence is that plastics should never become a waste eventually ending up in marine waters. Thus, they should be given no chance to disintegrate into microplastics, pollute oceans and terrestrial environments, and become a health hazard.

9.7 CONCLUSIONS AND FUTURE OUTLOOK

Marine ecosystems are complex, dynamic, and are the largest ones on the living planet. The health of the oceans and the seas are vital, with global implications for all life on Earth. All the aquatic ecosystems as well as all terrestrial ecosystems of the Earth are stressed by ever-increasing pollution. Plastic pollution is unabatedly turning the marine ecosystem condition grimmer, posing a threat to public health. Management of plastic waste and mitigation of microplastic pollution is not only necessary but an imperative of our times. To address this malignant state of the marine ecosystem, a global legislature system needs to be strengthened and made more effective and target-oriented. Implementation of various projects and programs at global, regional, and local levels for addressing various dimensions of multiple problems emerging from the "plastisphere" in the oceans is of critical importance and this strategy, in fact, has been widely recognized.

Plastic pollution control measures are being effectively managed in developed countries rather than in the developing ones. If such gaps in control measures remain a reality, the problem cannot be addressed at global scale to an appreciable extent. There should be no dearth of capacity-building mechanisms at any level and the developing countries' participation must be overwhelming, as a large chunk of the plastics emerges from these countries. EPR, one of the most effective concepts of plastic pollution control, needs to be strengthened in the developing countries that have not given adequate value to the environment cleanliness programs under this concept. The concept at the country level should extend down to village levels.

Eco-friendly management of plastic wastes and environmentally sound plastic manufacturing processes need to embrace all dimensions of ecological conservation including their impact on carbon sequestration enhancement and some degree of contribution to climate change mitigation. That should precisely be the state of the future "world of the plastics".

Scientific studies looking into the severity of the problem and evolving innovative technologies to effectively mitigate the problem will be instrumental in achieving desirable outcomes. The political strategic solution of the problem lies in the implementation of environmental governance at every level—ranging from local to global. In this system of governance, maintenance of the quality of our environment to be ensured by all desirable measures should be the central mandate of the governance bodies. Lifestyles based on the philosophy of 4 Rs—reduce, reuse, replace, and recycle of all usable plastic products—is the ultimate solution.

9.8 SUMMARY

Growing public demand is driving the world's governments and international organizations to address macro- and microplastic related issues through the development of new legislative measures, sustainability initiatives, and regulations. Sustainable Development Goal 14 (SDG 14), one of the 17 SDGs propounded by the United Nations General Assembly in 2015, focuses on "conserve and sustainably use oceans, seas and marine resources for sustainable development". The United Nations Convention on the Law of the Sea (UNCLOS) adopted in 1982 contains two articles—Article 207 and 211—that have focus on the concern relating to reduction, control, and prevention of plastic litter in marine ecosystems. The UN resolutions seek comprehensive and integrated efforts of international, regional, and national organizations towards finding solutions for marine pollution arising due to increasing plastic litter accumulation. The prominent international organizations include UNEP, FAO, and IMO. These are assisted by sub-regional fisheries management organizations. The UN Convention on Sustainable Development (Rio +20) highlighted the urgency of plastics pollution control in ocean basins. MARPOL is the main legislature's body to look into the objectives set out for marine pollution management, including plastic pollution control and prevention. The Convention on the Conservation of Migratory Species (CMS), or the Bonn Convention—a United Nations Environmental Treaty—is assigned many tasks relating to plastic pollution control and prevention.

UNEP's Global Partnership on Waste Management (GPWM) has paved a path to waste management in collaboration with international organizations and NGOs. MSFD of the European Union is aimed at reducing marine wastes to the minimum at a regional level. NOAA is looking after the Hawaii and South African coasts for waste management. In the new REACH regulations, microplastics are currently an issue of chemical risk management. Earlier, REACH regulations embraced the principle of "no data—no market" in a bid for safe use of chemicals. Additives as small molecules used in polymers as per REACH registration need to be tested against their deleterious effects on human health and the environment. All the plastic products falling under the ECHA's broad microplastic definition would need to be labeled and their uses to be reported to ECHA. Natural, soluble, and biodegradable polymers, however, are outside the scope of any regulation. The plastic products that do not lose their particle nature through processing are to be banned.

The first major regulatory directions by India emanate from the Environment (Protection) Act of 1986. India's first rule regarding plastic wastes was the Recycled Plastics (Manufacture and Usage) Rules 1999. The Bureau of Indian Standards (BIS) is the body that issues necessary regulations pertaining to plastic, polyethylene PET, and PET recycle usage. The Plastics (Manufacture, Usage and Waste Management) Rules 2009 were passed subsequent to those of 1999 to regulate plastic wastes more effectively. Plastic Waste (Management and Handling) Rules 2011 was the next ruling. The Government of India brought to the light the Plastic Waste Management Rules in 2016 which were further amended in 2018. The major responsibilities of implementing the rules have been assigned to Central Pollution Control Board (CPCB) and all the State Pollution Control Boards (SPCBs).

The consumers-citizens' role in reducing plastic pollution is phenomenal. Most countries have laws to control pollution created by industries. The permits include categorical provisions about waste management, bans on hazardous plastic products/materials, and incentives for effectively adopting recycling practices. The European countries, for instance, have formulated legislation for the plastic regulation on production, transport, and pellets usage by means of EU Packaging Directive (Directive 2008/98/EC, 2008), REACH (Regulation (EC) No 1907/2006), and the Industrial Emissions Directive (Directive 2008/98/EC). Taxes need be imposed on the basis of indirect environmental damage of the products or on how recyclable the components of the products are. A right-based action used for targeting the plastic industry is what is known as Extended Producer Responsibility (EPR). The EPR concept attempts to address the responsibilities towards a greener and cleaner environment even after the production chain is completed. If the industrial behavior is antagonistic to the environment and consequently to public health, consumers are driven to change their behavior.

One of the approaches to handle the problem more effectively is to set priorities and change plastic regulations based on innovations and relative impact on the marine environment. Use of atmospheric CO_2 as a feedstock in the process of PUR plastic production, possible due to advances in catalytic science, would be a revolutionary approach. Three tons of atmospheric CO_2 are utilized (removed from the atmosphere) for each ton of epoxide replaced with CO_2 as a raw material. The plastic regulation bodies must provide incentives to the firms that switch over to the alternative environment-friendly processes of plastic production. Applications of nano-engineering for the purpose of creating recyclable materials enabling replacement of complex non-recyclable multi-layered packaging may be brought into use. A magnetic additive can create air and moisture insulation effectively and thus save food items—especially some sensitive ones including coffee and medicines—from getting spoilt. Biodegradable and compostable agriculture and forestry by-products could be environment-friendly alternatives of the conventional petroleum-based plastic packaging material. Innovations in plastic wastes management and microplastic pollution mitigation need be incorporated into the principles of a circular economy.

Scientific studies looking into the severity of the problem and evolving innovative technologies to effectively mitigate the problem will be instrumental in achieving desirable outcomes. Environmental governance at every level—ranging from local to global level—would be a promising political development. Lifestyles based on the philosophy of 4 Rs—reduce, reuse, replace, and recycle all usable plastic products—is the ultimate solution.

REFERENCES

Alpizar, F., Carlssonic, F., Lanzab, G., Carneyd, B., Danielse, R. C., Jaimef, M., Hog, T., Niec, Z., Salazarh, C., Tibesigwai, B. and Whaderaj, S. 2020. A framework for selecting and designing policies to reduce marine plastic pollution in developing countries. *Environmental Science and Policy*, 109:25–35. doi: 10.1016/j.envsci.2020.04.007.

Backhaus, T. and Wagner, M. 2019. Microplastics in the environment: Much ado about nothing? *Global Challenges*, 4:1900022. doi: 10.1002/gch2.201900022.

Blackburn, A. 2018. Why innovation around plastic can benefit the environment? *European Journal of Environment and Earth Sciences*. https://environmentjournal.online/articles/why-innovation-around-plastic-can-benefit-the-environment/

Boucher, J. and Friot, D. 2017. *Primary Microplastics in the Oceans: A Global Evaluation of Sources*. Gland: International Union for Conservation of Nature and Natural Resources (IUCN). 43pp. doi: 10.2305/IUCN.CH.2017.01.en.

Brouillat, E. and Oltra, V. 2012. Extended producer responsibility instruments and innovation in ecodesign: An exploration through a simulation model. *Ecological Economics, Sustainability in Global Product Chains*, 83:236–245. doi: 10.1016/j.ecolecon.2012.07.007.

ECETOC (European Centre for Ecotoxicology and Toxicology of Chemicals). 2019. *Conceptual Framework for Polymer Risk Assessment (CF4Polymers)*. Brussels, Belgium: ECETOC Technical Report 133–1. www.ecetoc.org/wp-content/uploads/2019/06/ECE-TOC-TR133-1CF4Polymers.pdf.

Gilli, M., Mancinelli, S. and Nicolli, F. 2018. Introduction. In: Gilli, M., Mancinelli, S. and Nicolli, F. (Eds.), *Household Waste Management: Some Insights from Behavioural Economics*. Cham: Springer International Publishing. 1–4. doi: 10.1007/978-3-319-97810-9_1.

Iles, J. 2018. 5 innovations that could end plastic waste. *GreenBiz*. www.greenbiz.com/article/5-innovations-could-end-plastic-waste

Marazzi, L., Loiselle, S., Anderson, L. G., Rocliffe, S. and Winton, D. J. 2020. Consumer-based actions to reduce plastic pollution in rivers: A multi-criteria decision analysis approach. *PLoS ONE*, 15(8):e0236410. doi: 10.1371/journal.pone.0236410.

Mitrano, D. M. and Wohlleben, W. 2020. Microplastic regulation should be more precise to incentivize both innovation and environmental safety. *Nature Communications*, 11:5324. doi: 10.1038/s41467-020-19069-1.

Plastics for Change. 2020. What are the regulations for the use of recycled plastic? www.plasticsforchange.org/blog/category/what-are-the-current-regulations-for-the-use-of-recycled-plastic.

Ryberg, M. W., Hauschild, M. Z., Wang, F., Averous-Monnery, S. and Laurent, A. 2019. A global environmental losses of plastics across their value chains. *Resources, Conservation and Recycling*, 151:104459. doi: 10.1016/j.resconrec.2019.104459.

Thushari, G. G. N. and Senevirathna, J. D. M. 2020. Plastic pollution in the marine environment. *Heliyon*, 6(8): e04709. Doi: 10.1016/j.heliyon.2020.e04709.

UNEP (United Nations Environment Program). 2011. *Towards a Green Economy: Part II Waste, Investing in Energy and Resource Efficiency*. 632pp. http://all62.jp/ecoacademy/images/15/green_economy_report.pdf

UNEP (United Nations Environment Programme) 2018. *Mapping of Global Plastic Value Chain and Plastic Losses to the Environment (with Particular Focus on Marine Environment)*. Nairobi: United Nations Environment Programme. 96pp. www.unep.org/.

United Nations. 2017. Resolution adopted by the General Assembly on 6 July 2017. Work of the Statistical Commission pertaining to the 2030 Agenda for Sustainable Development (A/RES/71/313). https://documents-dds-ny.un.org/doc/UNDOC/GEN/N17/207/63/PDF/N1720763.pdf?OpenElement

Walker, T. R. 2021. (Micro)plastics and the UN Sustainable Goals. *Current Opinion in Green and Sustainable Chemistry*, 30:100497. doi: 10.1016/j.cogsc.2021.100497.

10 Role of Community Participation for the Reduction of Microplastic Emissions

Desirable changes in people's perceptions and behavior will be phenomenal towards addressing the issues relating to microplastic pollution. A community is the ultimate cause and the sufferer of the microplastic pollution, but also an ultimate source to bring substantial reduction in microplastic emissions. Education, research, NGOs, media and social networks, etc. can be instrumental towards catalyzing and strengthening community participation to deal with the problem. Reducing plastic consumption, improvement in production efficiency and recyclability, avoiding production of hazardous and certain single-use plastics, increasing products' usability age, etc. are parts of the solution. Development and implementation of an Integrated Waste Management System (IWMS) and fixing corporate responsibilities like Extended Producer Responsibility (EPR) are helpful towards reducing microplastic emissions. Community lifestyles encompassing the 5Rs (reduce, reuse, replace, recycle, and recover) and use of plastic products within a circular economy would be phenomenal for environmental conservation—including microplastic pollution control—through meaningful community participation.

Unimaginable proliferation in the use of plastic products in recent decades is attributable to a mix of socioeconomic factors. Social perception about an artificial product eventually determines the rates of its production. A huge variety of plastics has come into being due to increasing public demand owing to the overwhelming use of plastic products in almost all walks of human life. Plastics—in essence—are central to modern human lifestyles. This "plastic culture" of the modern era is leaving behind very large plastic footprints. It is true that the general public is responsible for plastic pollution extending up to its worst form—the microplastics pollution in the marine environment. But it is also true that desirable changes in human perceptions and behavior will be phenomenal towards addressing the microplastic issues.

10.1 WHY IS REDUCTION OF MICROPLASTIC EMISSIONS IMPORTANT?

Human activities dictated by over-consumptive behavior emanating from modern lifestyles have led to the creation of a huge "plastisphere" extending from the highs of the Himalayas to the depth of the oceans. Antagonistic to the biosphere, this

DOI: 10.1201/9781003312086-10

"plastisphere" is posing a threat to the planet's own life in one way or the other, and challenging its own "creator"—the human species! All human development dimensions—physical, intellectual, ethical, and aesthetic—are being badly affected by the proliferating plastic pollution.

Owing to the grave and far-reaching implications of the microplastic pollution, significant reduction and eventually mitigation of the microplastics pollution has become an imperative for humanity in our times. Ill effects of microplastics ranging from those on marine biota to those on human health are staggering and call for urgent attention and implementation of various measures. Ubiquity of microplastics makes us become especially cautious about the pollution. The most serious aspect of the microplastics pollution is the fact that oceans and seas are the ultimate home to the microplastics. Since the pollution itself is a dynamic process and knows no ecosystem boundaries, it is bound to infest terrestrial life along with the marine life—through ecosystem linkages, including food chains.

Impacts of macroplastics and microplastics on the environment are now well understood. Human demands to avail life's essentials, for instance food, energy, houses, etc., rest on the resource base already under stress due to macroplastics and microplastic pollution. Ugly scenarios of environment disruption are reflected in plastic litter everywhere and microplastic loads in the marine environment. Carrying capacity, or capacity to fulfill basic human demands, of the plastic-stressed environment might diminish and eventually be exhausted with the constantly increasing plastic pollution. The greater the intensity of the macro- and microplastic pollution, the lower the carrying capacity of the affected ecosystem. Environmental disruption induced by macro- and microplastics negatively affects human welfare. The human response fulfilling essential demands from the plastic-shrunk resource base and deteriorated environment, thus operates with a vicious cycle (Figure 10.1). Such a state of being itself creates a demand to resolve the issues relating to the microplastics.

Humans, undoubtedly, have capabilities to reduce the size of the "plastisphere" and even erase the plastic footprints. And translating the microplastic reduction and controlling microplastic pollution into a reality through the applications of various tactics is one of the pressing demands of our times.

10.2 HOW CAN COMMUNITY PARTICIPATION REDUCE MICROPLASTIC EMISSIONS?

Status of the production and marketing of a product of socioeconomic interest depends on public demand. Plastic items of huge varieties have been and continue to be in use in day-to-day life and, thus, are an object of overwhelming public demand. Meeting the economics underlying the "demand and supply" principle, the plastic products are pouring into the global market at accelerated rates. The plastic products fetch huge profits and no dearth of the raw material prompts the plastic industry to go on manufacturing the items presumably at an alarming rate. This kind of prevailing atmosphere paves the path for the constant generation of plastic wastes and microplastic emissions. Thus, in essence, the public is the ultimate cause and the sufferer

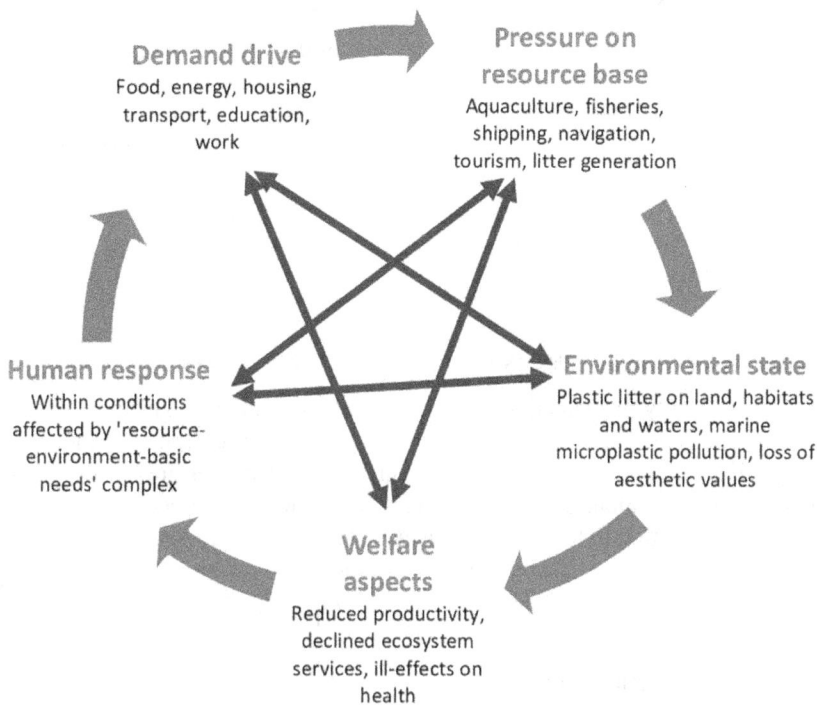

FIGURE 10.1   The vicious cycle created by macro- and microplastics.

of the microplastic pollution. Contrarily, the public would also be the ultimate source to bring about the substantial reduction of microplastic emissions.

Community action in our world is the most pivotal. No transformation in any area can be possible without community participation. Many examples of community action in our times have glorified many communities themselves and their leaders. The Hug-the-Trees Movement (*Chipko Andolan*), Save Seed Movement (*Beej Bachao Andolan*), Save Ganga-Himalaya Movement (*Ganga-Himalaya Bachao Andolan*), Save Narmada Movement (*Narmada Bachao Andolan*), etc. are the notable community movements in India that have been successful to a great extent. In many programs of ecological development, governments generally seek people's participation. Large-scale afforestation to develop community forests through community participation is another excellent example in India.

In the various legislations chalked out and in programs relating to plastic debris management, participation of various communities at the local level (city, town, village levels—for instance) should be duly ensured. There could be three dimensions of community participation: (i) creating awareness about the harmful effects of microplastics; (ii) direct action for waste collection and proper disposal; and (iii) exerting influence on the plastic industry and plastic regulatory bodies and governments for strictly adopting concrete measures for microplastic reduction and developing innovations in the processes of plastic material production.

10.2.1   REDUCING THE CONSUMPTION OF PLASTIC

Over-consumption of plastic material is one of the characteristics of modern society—a fact that culminates into microplastic pollution. The community can play a crucial role in reducing plastic consumption, the first step towards controlling microplastic pollution. The various ways a community can adopt plastic reduction are (i) creating awareness about the deleterious effects of microplastic pollution on human health and on the environment; (ii) replacing plastic items with non-plastic ones for which there are several alternatives; and (iii) adopting frugal lifestyles—with grace without waste.

10.2.2   RAISING AWARENESS IN PUBLIC THROUGH EDUCATION

Lack in community participation or a slow pace of community participation is primarily on account of the lack of awareness about the causes and consequences of macroplastics and microplastic pollution. Further, communities are hardly aware of various regulations about pollution control and laws/legislation necessary to reduce microplastic pollution. There is no dearth of the knowledge about all aspects of macro- and microplastics. A community can embark on raising awareness in public through education that would be instrumental in reducing microplastic pollution.

10.2.3   ROLE OF THE RESEARCH

Research on various aspects and issues relating to a variety of polymers in general and on microplastics and nanoplastics in particular has generated enormous information about all the dimensions and consequences of the artificial polymers. Severe effects of the polymers fragmented into micro- and nanoplastics have been investigated through research. Gradation of plastics and associated additives and other chemicals on the basis of their harmful effects on human health has been established through research. Further, methodologies of plastic biodegradation using various strains of microorganisms, especially bacteria and fungi, have been recorded in the published research papers. Mechanisms on plastic recycling are well established. Nowadays research focus is on evolving innovative techniques for manufacturing more eco-friendly polymer products, gainful management of plastic wastes, prevention of microplastic emissions, and on more effective mitigation of microplastic pollution in marine environments. Such research findings provide practical ways of dealing with the growing "world of plastics".

10.2.4   ROLE OF NGOS

Non-government organizations (NGOs) have been playing a very important role in environmental and ecological conservation across the world. The NGOs can also play a pivotal role in catalyzing and strengthening community participation for reducing microplastic emissions. NGOs, unlike government-run organizations, are often more sensitive to socio-cultural, ecological, and environmental issues. NGOs are generally constituted based on unique local/community-related issues and their objectives and activities are also based on communities' aspirations. An enhanced role of NGOs in the plastic regulatory systems would be phenomenal for effective community participation for reducing microplastic emissions.

10.2.5 ROLE OF THE MEDIA AND SOCIAL NETWORKS

Media and social networks are ruling over the Information Age the contemporary world is living in. Media and social networks have practically occupied the center stage of modern communication systems. No task, no program, no project, no regulatory system, no development, and no governance are practically functioning without the overwhelming use of media and social networks. Use of media and social networks, in essence, is inevitable for ensuring and efficiently enhancing community participation towards reducing microplastic emissions, plastic waste management and evolving more promising plastic regulatory systems.

10.3 IMPROVING PRODUCTION EFFICIENCY OF PLASTIC PRODUCTS

Plastic's entry into almost all walks of human life poses unprecedented challenges. The first and the foremost way of meeting this challenge is to reduce plastics at the production level itself. It can be accomplished by:

1. Using alternatives to plastics, such as glass, aluminum, plant-based products, biodegradable materials, etc.
2. Improving manufacturing designs that could substantially decrease the amounts of plastics used.
3. Increasing usability age of plastic products by repair and prolonged use of good quality products.
4. Avoiding production of certain kinds of single-use plastics, many of which have been banned.
5. Improving recyclability of unusable plastics/plastic wastes by restricting the numbers of polymers, additives, and mixers.

Substantial improvement in designs not only contributes to improve production efficiency of plastic products, but is also beneficial for the plastic industry due to decreased demand for raw material. The plastic products produced using recycled plastic material are more expensive than those made of virgin plastic (Prata et al. 2019). However, recyclability of the unusable plastic products/wastes is an environment-friendly process as it prevents waste generation and subsequent fragmentation into microplastics upon entering into marine ecosystems and/or freshwater environments. Being beneficial from an environment point of view means being correspondingly beneficial from a societal point of view. Recyclability, thus, needs to be encouraged as a marketing strategy (Walker and Xanthos 2018) and a community of consumers should be apprised of the benefits from it.

Some portions in some usable products and some plastic items as such may have no simple plastic substitutes, yet most of them may be substituted by biodegradable substitutes. The plastic part in cotton buds can be substituted by paper. Microbeads can be substituted by biodegradable material. A total ban on single-use plastics is always promising. Such a ban has already been imposed in Canada, New Zealand, USA, and United Kingdom (Dauvergne 2018). Plastics like polycarbonate, polyvinyle chloride, polystyrene, and polyurethane need to be classified as hazardous (Rochman and Browne 2013). A ban on selective plastic types should be on the basis of their impact on the environment and overall life. Ban on account of the direct impact on human health should not be a singled-out criterion. Human health is not just an attribute of itself. The status of health phenomenally depends on the quality of environment. If the microplastics and/or the additives used in plastic items impact lower forms of aquatic or terrestrial life, humans cannot refrain from being affected through food chain. It would not be appropriate to suggest a ban on the plastics used in medical equipment and in the instruments of critical needs. However, improvement in the quality of the plastic used—to the extent that it is safe for human health—is necessary. It must be mandatory for the plastic industry to confirm that the plastics they use in designing varieties of items for human use is safe for the environment and public health.

Plastic manufacturing firms urgently need to reduce their production and increase recycling rates during manufacturing and for this to work, they should to rely on voluntary and mandatory measures. Again, the firms must also be accountable for the waste generated by the products they produce and market under the Extended Producer Responsibility (EPR) that can be translated into reality by paying a recycling fee (Prata et al. 2019). The EPR, in fact, would be vital for reducing wastes, enhancing support for recycling, and increasing efficiency. Meeting ecological requirements, environmental safety norms, labeling of hazardous/harmful substances by the plastic industry should be mandatory. Incentives to the firms sincerely implementing these conditions would be helpful towards controlling and mitigating the problem arising due to plastic production.

10.4   REDUCING MICROPLASTICS THROUGH PROPER DISPOSAL OF PLASTIC WASTE

Microplastics pollution in the marine environment is rooted in the improper disposal of plastic waste. Proper disposal of the plastic waste, therefore, would be phenomenal in significantly reducing microplastics. In the "journey" of plastic products from production to consumption, plastic waste generation is inevitable. Intervention in this "journey" for determining the products' fate after its use, therefore, is necessary for reducing and managing waste and avoiding consequent problems. Rampant generation of plastic waste, however, becomes problematic, as the whole world is already experiencing. This problem arises out of sheer negligence of plastic wastes and lack of proper waste disposal systems. A proper plastic waste disposal system, therefore, is one of the keys to controlling microplastic pollution.

Schneider and Ragossnig (2015) have suggested an integrated approach to waste management with recycling as a core component, and the same could be useful for operationalizing the waste-to-energy concept. In consonance with the same, let us suggest an Integrated Waste Management System (IWMS) for effective, environment-friendly, and socioeconomically useful plastic waste disposal. Reuse of plastic packaging is somewhat difficult due to recovery and sorting requirements. Therefore it is hardly used outside the high-value goods (Liu et al. 2018).

Therefore, more appropriate approaches for the generated waste are (i) its being recycled, (ii) its use in feedlot and energy recovery (only when it is not to be recycled), and (iii) landfilling of the final waste (e.g., ash). Such a plastic waste disposal approach would result in decreased plastics in the environment and consequent fragmentation into microplastics. The Plastic Restriction Policy and Recycling Act and Compulsory Trash-Sorting Policy adopted in Taiwan have been instrumental in reducing per capita plastic waste disposal rates to the extent of about 50% (Brennholt et al. 2018). The other example is from Australia where, due to higher investment in waste management, significant reduction in plastic wastes on coasts was recorded (Willis et al. 2018). With over 180 countries as the signatories, the Basel Convention has the regulations for the trade of mixed plastic scrap between countries (Prata et al. 2019).

The IWMS, however, is difficult to be implemented due to high investment. Again it is a somewhat slow process not to yield quick results. In case such integrated system is not operative, such as in poor countries, incinerators and landfills can be used successfully until the IWMS is developed. International organizations must come forward to provide financial and technical aids towards the development and implementation of IWMS. Local governance, the public-private sector, and community participation can play crucial roles towards operationalizing IWMS, which, in fact, is a global need.

10.5 RECYCLING

Recycling of plastic waste is of absolute importance for sound environment-friendly management and also as a part of a circular economy. Recycling, however, is a very complex, energy-intensive, and quite expensive process involving several steps as shown in Figure 10.2. In this process, the separation by polymer is pretty difficult and it is due to this that the quality of the recycled product is likely to be compromised.

Recycling may be of two types: primary recycling (closed-loop) and secondary recycling (downgrading). The primary recycling generates high-quality plastics using uncontaminated materials. This is the plastic produced by manufacturers. The examples are back covers of plastic on flat screen TVs. The secondary recycling, on the other hand, contributes to generate lower-quality plastic material generally used in applications comparatively in lower demand. The examples of the secondary recycling are asphalt, concrete, consumer waste, composites, and construction materials (Poulikakos et al. 2017).

There are many reasons why recycling of the plastic left as waste after consumers' use is going on at a slow pace. These are (i) higher costs of the recycling process than that of virgin plastic; (ii) limited use of the recycled products owing to declined quality due to contamination, including organic contamination; (iii) reduced number of recycling cycles; (iv) difficulties in constant supplies of standard quality raw material to the manufacturing firms; and (v) lower recyclability of some of the waste originating from plastic items like textiles, and laminated plastic material.

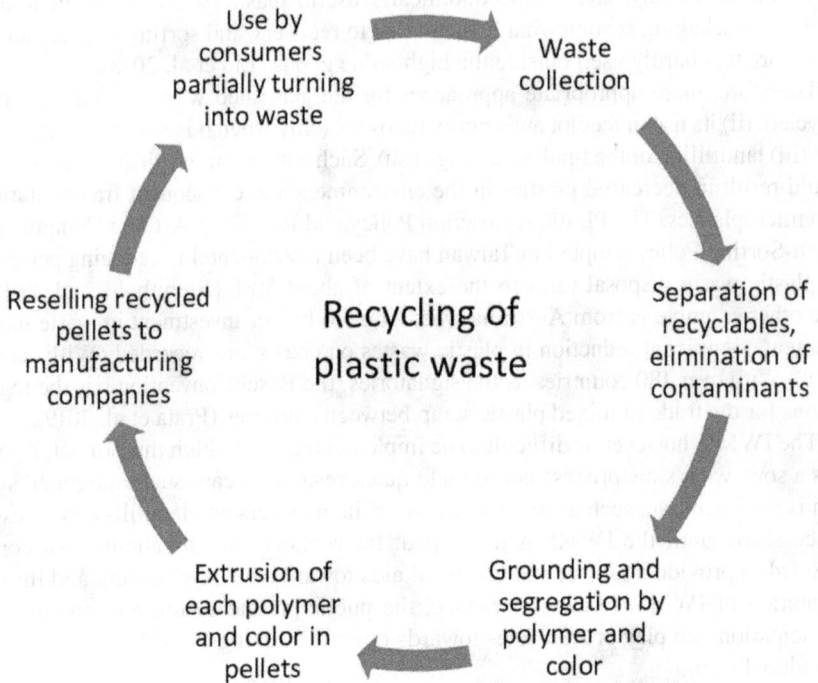

FIGURE 10.2 Various steps in the process of plastic recycling.

Recycling of plastics may not always be economically feasible. However, it may help regenerate economic returns in closed-loop systems (Bernardo et al. 2016). Some of the plastics, especially polyethylene terephthalate and polyethylene, necessitate only half the energy to produce virgin polymers (Arena et al. 2003). Incineration is likely to be favored when organic contamination is high and probability of replacing virgin plastics is low (Lazarevic et al. 2010). Recycling rates can, undoubtedly, be increased to an appreciable extent provided constant supplies of the raw material of standard quality are maintained and technological developments leading to enhanced recycling efficiency take place.

Negative effects of the recycling process on the environment always emanate from the non-renewable energy source consumed in the process and in transportation and due to the fillers and additives introduced into the recycled product (Arena et al. 2003; Lazarevic et al. 2010; Gu et al. 2017). Despite high costs and many difficulties faced and some negative effects on the environment, the overall positive environmental and socioeconomic attributes of plastic recycling make it a preferable technique of waste management. When recycling of the plastic wastes goes on, no or very little land for refilling is required. The saved land can be utilized for purposes like ecological conservation and socioeconomic use. This process generates employment opportunities and becomes the basis for economic gains. After all, recycling is an integral component of the philosophy of the circular economy and sustainable development.

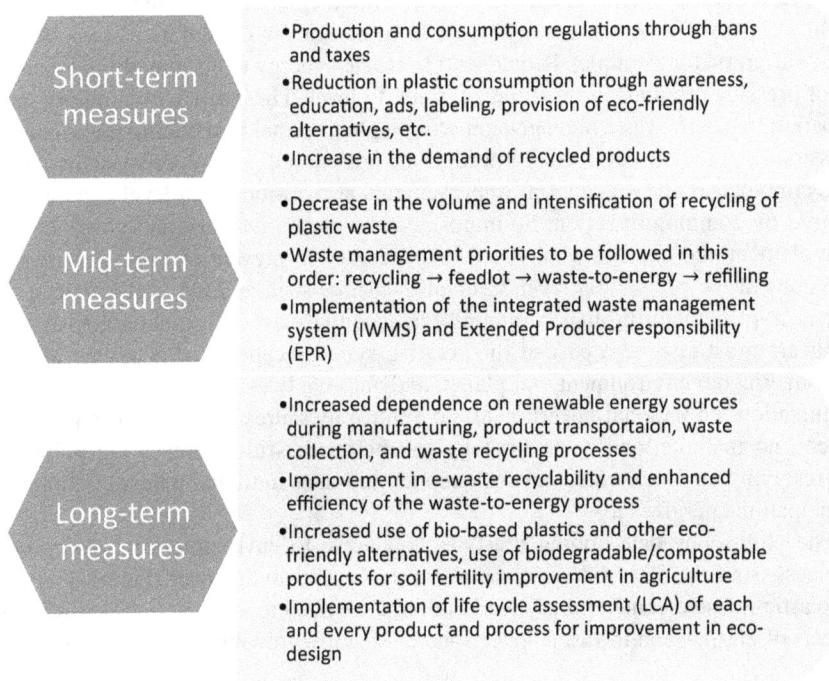

Short-term measures
- Production and consumption regulations through bans and taxes
- Reduction in plastic consumption through awareness, education, ads, labeling, provision of eco-friendly alternatives, etc.
- Increase in the demand of recycled products

Mid-term measures
- Decrease in the volume and intensification of recycling of plastic waste
- Waste management priorities to be followed in this order: recycling → feedlot → waste-to-energy → refilling
- Implementation of the integrated waste management system (IWMS) and Extended Producer responsibility (EPR)

Long-term measures
- Increased dependence on renewable energy sources during manufacturing, product transportaion, waste collection, and waste recycling processes
- Improvement in e-waste recyclability and enhanced efficiency of the waste-to-energy process
- Increased use of bio-based plastics and other eco-friendly alternatives, use of biodegradable/compostable products for soil fertility improvement in agriculture
- Implementation of life cycle assessment (LCA) of each and every product and process for improvement in eco-design

FIGURE 10.3 Some crucial and applicable short-, mid-, and long-term recommendations for reducing/preventing microplastic emissions during plastic production, consumption, and disposal.

10.6 RECOMMENDATIONS

Recommendations for reducing microplastic emissions during production and consumption of plastic products and disposal (management) of plastic wastes, classified into short-term, mid-term, and long-term measures, are depicted in Figure 10.3.

10.7 THE WAY AHEAD

All varieties of products being manufactured by various industries and available in the market are being used in human systems for a variety of purposes. Their demand arises as per their utility which is evaluated by the users/consumers from all angles. Plastics constitute one such matter that matters for the improved ways of living, but at the same time also spells out numerous problems for humanity as well as for the whole of life and for the global environment.

Accumulation of microplastics in the marine environment is going on rampantly thanks to overuse of plastic products, bottlenecks in production processes, and mismanagement of plastic wastes. This state of the planet's largest ecosystem is posing a threat to organisms, ecosystems, and human health. Since the microplastic pollution in the marine ecosystems is thanks to human activity—from production to waste generation—its solution also lies in large-scale people's participation right from policy formulation to appropriate waste management. People organize at community levels. People suffer at community levels (communities in coastal areas may be the worst sufferers, for example). People tend to create synergy to do something creative out of pressing circumstances at the community level. Thus, active community participation in plastic waste management will be phenomenal in reducing microplastic emissions and thus in controlling microplastic pollution in marine ecosystems.

Technological and managerial improvements at the production level can also be ensured by community action, by imposing pressure on industry and governments to implement and execute what could be necessary to prevent or minimize plastic emissions at the production level. Crucial measures such as LCA and EPR can be strictly implemented through community action. Plastic industry technologists and engineers must also be a part of the creative synergy capable of resolving all the issues facing our environment, our planet, and our own lives. In fact, no international organization, no national government, no control measures, no laws, no regulatory system, no institute, and no technology can be successful in preventing pollution and resolving environmental and socioeconomic issues until and unless community participation energizes it.

The philosophy determining lifestyles conducive to environmental conservation revolves round the 5Rs: reduce, reuse, replace, recycle, and recover (Figure 10.4). In the plastic-infested contemporary world of ours, we need to strategically deal with all aspects of plastics and thwart a "plastisphere" steadily growing within the biosphere.

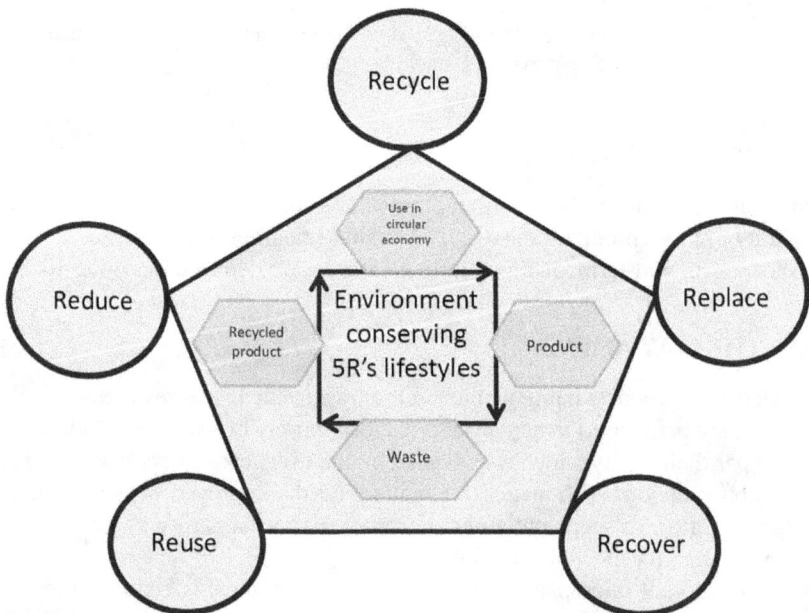

FIGURE 10.4 Community living philosophy based on the 5Rs with the use of plastic products within a circular economy. This would be phenomenal for environmental conservation—including microplastic pollution control—through meaningful community participation.

We should reuse the quality plastic products as long as possible which is necessary to avoid adding more plastic material for the same purpose and consequently increasing the volume of the waste material. Wherever possible, we can replace the plastic items by biodegradable and environment-friendly ones. Recycling of the plastic wastes prevents the unusable material ending up in lakes, seas, and oceans and getting fragmented into microplastics, and this eco-friendly process also generates employment and economic avenues for the people. With intensive plastic waste recycling using renewable energy sources, we shall be promoting a circular economy that takes care of the environment as well as the economy (Figure 10.4). In the process of waste management, we must also recover energy and valuable chemicals that would further add to benefits accrued through 5R living philosophy and circular economy framework.

10.8 SUMMARY

People are responsible for microplastics pollution in the marine environment to a large extent. Therefore, desirable changes in human perceptions and behavior will be phenomenal towards addressing the issues relating to this pollution. Owing to the grave and far-reaching implications of the microplastic pollution, acquiring significant reduction and eventually mitigation of the microplastics pollution are absolutely necessary in our times. Since the pollution itself is a dynamic process and knows no ecosystem boundaries, it is bound to infest terrestrial life along with the marine life—through ecosystem linkages, including food chains.

The greater the intensity of the macro- and microplastic pollution the lower the carrying capacity of the affected ecosystem. The public is the ultimate cause and the sufferer of the microplastic pollution. Contrarily, the public would also be the ultimate source to bring substantial reduction in microplastic emissions. No transformation in any area and no solution of any problem can be possible without community participation. There are three dimensions of community participation: creating awareness, direct action for waste collection and proper disposal, and exerting influence on the plastic industry, regulatory bodies, and governments to address all emerging problems.

Reducing plastic consumption, a community can play a crucial role and is the first step towards controlling microplastic pollution. Lack in community participation or slow pace of community participation is primarily on account of the lack of awareness about the causes and consequences of macroplastics and microplastic pollution. A community can embark on raising awareness through education that would be instrumental in reducing microplastic pollution. Research on various aspects and issues relating to a variety of polymers in general and on microplastics and nanoplastics in particular has generated enormous information about all the dimensions and consequences of the artificial polymers. NGOs can also play a pivotal role in strengthening and catalyzing community participation. Use of media and social networks is inevitable for ensuring and efficiently enhancing community participation towards reducing microplastic emissions, plastic waste management and evolving more promising plastic regulatory systems.

Improvements in production efficiency of plastic products can be accomplished by using alternatives to plastics, improving manufacturing designs, increasing usability age of plastic products, avoiding production of certain kinds of single-use plastics, and improving recyclability of unusable plastics/plastic wastes. Substantial improvement in designs not only contributes to improve production efficiency of plastic products, but is also beneficial for the plastic industry due to decreased demand for raw material. Recyclability of the unusable plastic products/wastes is an environment-friendly process as it prevents waste generation and subsequent fragmentation into microplastics upon ending up in marine ecosystems and/or freshwater environments. It must be mandatory for the plastic industry to confirm that the plastics they use in designing varieties of items for human use is safe for the environment and public health.

Plastic manufacturing firms urgently need to reduce their production and increase recycling rates, and to ensure it, they need to rely on voluntary and mandatory measures. Again, the firms must also be accountable for the waste generated by the products they produce and market under the Extended Producer Responsibility (EPR) that can be translated into reality by paying a recycling fee. Microplastic pollution in the marine environment is rooted in the improper disposal of plastic waste. Proper disposal of the plastic waste, therefore, would be phenomenal in significant reduction in microplastics. A proper plastic waste disposal system, therefore, is one of the keys to control microplastic pollution. Let us suggest an Integrated Waste Management System (IWMS) for effective, environment-friendly, and socioeconomically useful plastic waste disposal. Such a plastic waste disposal approach would result in decreased plastics in the environment and consequent fragmentation into microplastics. Recycling of plastic waste, a sound environment-friendly approach and a part of a circular economy, also generates employment opportunities and becomes the basis for economic gains.

A community living philosophy based on the 5Rs (reduce, reuse, replace, recycle, and recover) with the use of plastic products within a circular economy would be phenomenal for environmental conservation—including microplastic pollution control—through meaningful community participation.

REFERENCES

Arena, U., Mastellone, M. L. and Prerugini, F. 2003. Life cycle assessment of a plastic packaging recycling system. *The International Journal of Life Cycle Assessment*, 8:92–98. doi: 10.1007/BF02978432.

Bernardo, C. A., Simões, C. L. and Pinto, L. M. C. 2016. Environmental and economic life cycle analysis of plastic waste management options: A review. In: Holzer, C. H. and Payer, M. (eds.) *Proceedings of the AIP Conference*, 21–25 September. Graz, Austria. 1779pp. doi: 10.1063/1.4965581.

Brennholt, N., Heß, M. and Reifferscheid, G. 2018. Freshwater Microplastics: Challenges for Regulation and Management. In: Wagner, M., Lambert, S. (eds.) *Freshwater Microplastics. The Handbook of Environmental Chemistry*, vol 58. Springer, Cham. pp. 239–272. https://doi.org/10.1007/978-3-319-61615-5_12

Dauvergne, P. 2018. The power of environmental norms: Marine plastic pollution and the politics of microbeads. *Environment Pollution*, 27:579–597. doi: 10.1080/09644016.2018.1449090.

Gu, F., Guo, J., Zhang, W., Summers, P. A. and Hall, P. 2017. From waste plastics to industrial raw materials: A life cycle assessment of mechanical plastic recycling practice based on a real-world case study. *Science of the Total Environment*, 601–602:1192–1207. doi: 10.1016/j.scitotenv.2017.05.278.

Lazarevic, D., Aoustin, E., Buclet, N. and Brandt, N. 2010. Plastic waste management in the context of a European recycling society: Comparing results and uncertainties in a life cycle perspective. *Resources, Conservation and Recycling*, 55:246–259. doi: 10.1016/j. resconrec.2010.09.014.

Liu, Z., Adams, M. and Walker, T. R. 2018. Are exports of recyclables from developed to developing countries waste pollution transfer or part of the global circular economy? *Resources, Conservation and Recycling*, 136:22–23. doi: 10.1016/j.resconrec.2018.04.005.

Poulikakos, L. D., Papadaskalopooulou, C., Hofko, B., Gschosser, F., Falchetto, A. C., Bueno, M., Arraigada, M., Sousa, J., Ruiz, R., Petit, C., Loizidou, M. and Partl, M. N. 2017. Harvesting the unexplored potential of European waste materials for road construction. *Resources, Conservation and Recycling*, 166:32–44. doi: 10.1016/j. resconrec.2016.09.008.

Prata, J. C., Patricio Silva, A. L., da Costa, J. P., Mouneyrac, C., Walker, T. R., Duarte, A. C. and Rocha-Santos, T. 2019. Solutions and integral strategies for the control and mitigation of plastic and microplastic pollution. *International Journal of Environmental Research and Public Health*, 16(13):2411. doi: 10.3390/ijerph16132411.

Rochman, C. M. and Browne, M. A. 2013. Classify plastic waste as hazardous. *Nature*, 494:269–271. doi: 10.1038/494169a.

Schneider, D. R. and Ragossnig, A. M. 2015. Recycling and incineration, contradiction or coexistence? *Waste Management and Research*, 33:693–695. doi: 10.1177/0734242X15593421.

Walker, T. R. and Xanthos, D. 2018. A call for Canada to move toward zero plastic waste by reducing and recycling single-use plastics. *Resources, Conservation and Recycling*, 133:99–100. doi: 10.1016/j.resconrec.2018.02.014.

Willis, K., Maureaud, C., Wilcox, C. and Hardesty, B. D. 2018. How successful are waste abatement campaigns and government policies at reducing plastic waste into the marine environment? *Marine Policy*, 96:243–249. doi: 10.1016/j.marpol.2017.11.037.

11 Recent Cutting-Edge Solutions to Prevent Microplastics Pollution

Microplastic pollution in the oceans is all set to multiply in the future and assume alarming proportions if concrete measures to address the problem are not adopted in a due course of time. By 2015, annual production of plastics had reached the total amount of the entire human population on the planet, according to Worm et al. (2017), and by 2050 the plastics present in the oceans would outweigh fish, according to the World Economic Forum (2016).

Plastic production, collection, and disposal are also enhancing climate change processes through greenhouse gas (GHG) emissions. According to an estimate, GHG emissions due to plastics are likely to reach 15% of the total carbon budget by 2050 (World Economic Forum 2016). The GHG emissions contributed by conventional plastics were 1.7 Gt of CO_2-equivalent (CO_2e) in 2015, which, with the persisting trend, are likely to formidably grow to 6.5 Gt CO_2e by 2050 (Zheng and Suh 2019).

Microplastic pollution in the marine ecosystem can be handled in several ways as we have discussed in previous chapters. The cutting-edge type solutions have in their folds the technologies developed in recent years that might be usable for targeting hotspots of ocean plastics pollution: the prevention and cleaning-up technologies (Schmaltz et al. 2020). If we could prevent plastics entering into waterways and collect the pollutants that have entered into water bodies, the process of larger plastic particles fragmenting into microplastics can be effectively tackled and microplastic pollution problem in the marine environment can be solved to a significant extent.

11.1 PREVENTION TECHNOLOGIES

Prevention of microplastic pollution is preferable over the collection of the pollutants in the aquatic environment. But as we can presume that prevention of microplastics in a world so overwhelmingly dependent on plastics is almost impossible, collection of the plastic pollutants becomes inevitable. Despite immediate global action towards implementing a circular economy, as much as 710 MMT of plastic waste is estimated to be released into the environment by 2040 (Lau et al. 2020).

Therefore, technologies for both, prevention and collection, are needed to be implemented simultaneously. Schmaltz et al. (2020) have made an inventory of as many as 52 technologies, most of them currently in application regarding prevention and collection of plastic pollution, out of which 59% are those applied for the collection of macroplastic waste already present in waterways.

DOI: 10.1201/9781003312086-11

All larger size plastics fragment into microplastics and nanoplastics and release deleterious additives in every environment. There is stray leakage of plastic particles at every stage of the plastic life cycle: pre-production (in the form of pellets), production, consumption, and waste management (recycling; reuse/repurposing, such as waste-to-energy; and landfilling). Innovative prevention technologies are especially needed to prevent these plastic leakages.

The prevention technologies have emerged from many types of organizations, such as governments, NGOs, academics, research and development (R&D), and corporations. In addition, entrepreneurs and experts have also been involved in the task of bringing in innovative technologies. Innovations happen at every level. No technology, in fact, can be the final one. Possibilities of renewed innovations are always there and innovators work everywhere, within organizations as well as among the experts and entrepreneurs working independently. Investment in the development of technologies is always promising and spells out numerous short-term and long-term attributes for the environment and human society. Influencers also catalyze the innovation processes in several ways. They could be among the users and from the public (Figure 11.1). The best criterion relating to innovation must include working efficiency, low-volume-high-value, cost-effectiveness, and users' accessibility.

Advanced technologies applied in WWTPs can potentially reduce microplastics discharged from the treatment plants. The membrane bioreactor (MBR) has been found to be quite an efficient technique with the demonstrated microplastic removal performance. This technology has the potential to remove 99.9% microplastics (Talvitie et al. 2017).

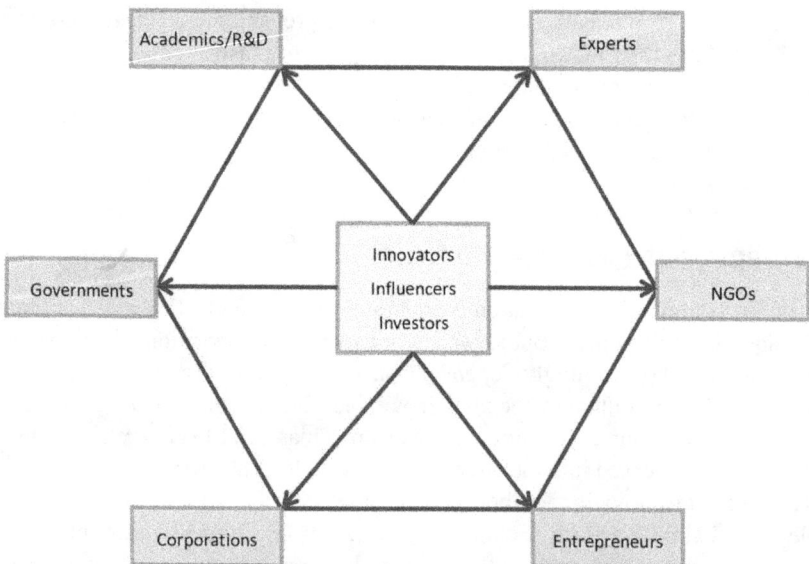

FIGURE 11.1 Different stakeholders contributing to the development of cutting-edge technologies addressing prevention and collection of plastic pollution.

The innovative technologies are applicable at various stages of the plastic cycle at which plastic pollutants' leakage is occurring, to prevent or minimize the leakage and contribute to averting or reducing microplastic pollution that eventually leads to marine pollution. As about 80% of the plastic pollution in the marine ecosystem is owing to land-based sources (Ritchie and Roser 2018), efficient prevention technologies applied at the pre-production, production, consumption, and waste management levels might be instrumental in reducing marine microplastic pollution. The UN Environment Assembly in its Resolution 2/11 appealed to the UN member countries for regional cooperation and worldwide cleanup actions and for development of environmentally sound systems and methods for removal and sound disposal of marine litters (United Nations Environment Assembly 2016). Such policy efforts floated and implemented by national governments have led to creating technologies worth preventing plastic wastes from making entry into the environment or collecting prevailing marine plastic pollutants.

The prevention technologies currently aimed at addressing leakage of plastic pollutants—macro- and microplastics—that are in operation are listed in Table 11.1.

A perusal of Table 11.1 reveals that the plastic pollution prevention technology targets both macroplastics and microplastics. Six technologies target microplastics and eight

TABLE 11.1
Plastic pollution prevention technologies.

Prevention technology	Name of the technology	Year of invention	Process/Functioning
Macroplastics			
Stormwater and wastewater filters	In-Line litter Separator (ILLS)	1999	Removes litter from passing stormwater
	Storm Trap Trash Trap	2018	Mesh net system uses water flow to capture and remove trash, floatables, and solids from storm- and wastewater
	Pump Guard	2016	Mesh nets remove debris from storm- and wastewater
	Watergoat Trash Trap	2006	Floating boom and net attached to embankments, stormwater outfalls, canals, or creeks collects floating debris
	Netting Trash Trap System	1999	Mesh nets capture and remove trash from stormwater and discharge
	StormX Netting Trash Trap	1995	Commercial grade reusable nets used to capture pollutants even of smaller size than 5 mm from stormwater runoff, including organic pollutants rich in nitrogen and phosphorus, thus helping reducing the level of nutrients responsible for algal bloom
Miscellaneous leakage prevention	Stow It, Don't Throw It	2012	Tennis ball containers repurposed into fishing line recycling bins for anglers
	CLEVER-Volume	2019	Sensors allow port authorities to certify the amount of ship waste reported, in comparison to the volume reported to the MARPOL inspectors

(Continued)

TABLE 11.1
(Continued)

Prevention technology	Name of the technology	Year of invention	Process/Functioning
Microplastics			
Laundry balls	Cora Ball	2019	Balls placed in the laundry machine capture microfibers shed when washing synthetic fibers
	Fiber Free	2017	Balls placed in the laundry machine or dryer capture microfibers shed when washing or drying synthetic fibers
Residential wastewater treatment	Lint LUV-R	2016	Water filter or laundry machines capture microfibers when water drains through the machine
Miscellaneous removal from wastewater	Unnamed invention by students of Gering High School, USA	2017	Gravity-fed, three-stage attachable filter catches microplastics such as microfibers shed from the synthetic clothes in laundry before they enter into wastewater
	GoJelly Project	2018	A gelatinous solution to microplastic pollution; jellyfish mucus secreted at the time of reproduction and under stress conditions captures and binds to nano-sized particles removing microplastics from wastewater
	Showerloop	2012	Filter removes microplastics while primarily filtering water for reuse

Source: Based on the inventory made by Schmaltz et al. (2020).

macroplastics. Among the prevention technologies, currently only nine technologies are in use. Two of the prevention technologies, Cora Ball and Lint LUV-R, are in use in USA and Canada respectively, and use of the Showerloop technology is unknown.

There are six kinds of technologies preventing the leakage of microplastics, with two technologies in each of the three categories. The two types of the prevention technologies targeting macroplastics have eight different brand names. On the whole, the "stormwater and wastewater filters" category has the largest number of technologies (i.e., six) (Figure 11.2). The technologies drastically reducing plastic leakage into natural systems have been proven in the lab, in pilot projects, and at industrial scale (World Economic Forum 2016).

11.2 COLLECTION TECHNOLOGIES

Collection technologies currently in operation targeting macroplastics, microplastics, and both, are listed in Table 11.2.

In the list of the macro- and microplastic collection technologies currently in operation (Table 11.2 and summarized in Figure 11.3), 31 are largely applicable for macroplastics, three exclusively for microplastics and four are applied for both macro- and microplastics. Of course, all the collection technologies would eventually take care of the microplastic pollution as the macroplastics would eventually fragment into microplastics in the marine environment. Among them all, the six collection technologies are not generally in use and three are under pilot testing (Schmaltz et al. 2020).

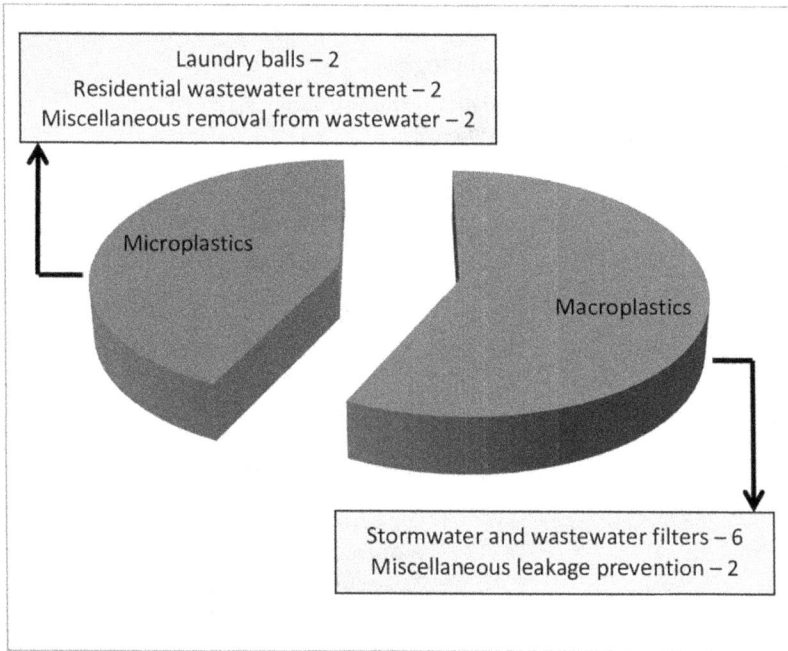

FIGURE 11.2 The type and number of prevention technologies targeting macro- and microplastics. Source: Based on the information documented by Schmaltz et al. (2020).

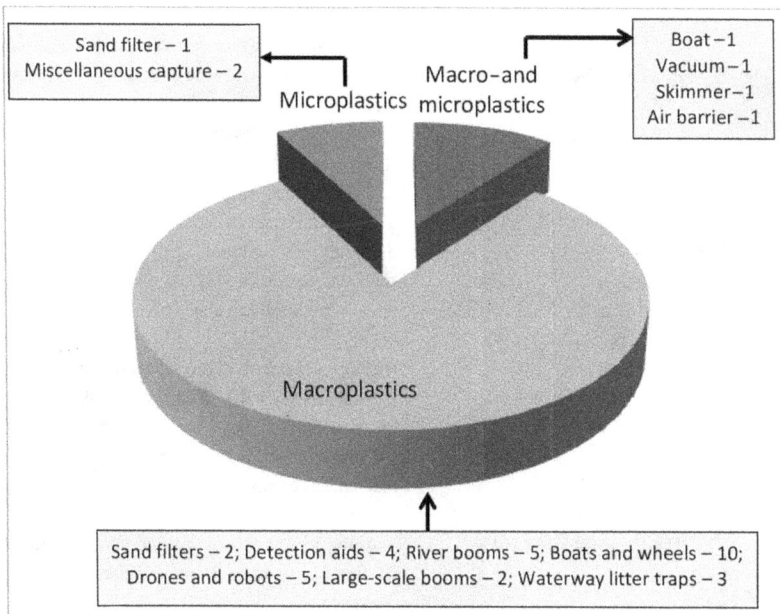

FIGURE 11.3 The type and number of collection technologies targeting macro- and microplastics. Source: Based on the information compiled by Schmaltz et al. (2020).

TABLE 11.2
Plastic pollution collection technologies.

Prevention technology	Name of the technology	Year of invention	Process/Functioning
Macroplastics			
Large-scale booms	Ocean Clean-Up System	2013	C-shaped boom and screens use currents to corral trash
	Holy Turtle	2018	1000-foot-long floating unit is towed by two marine vessels and captures floating waste; large vent hole protects marine life
Drones and robots	FRED (Floating Robot for Eliminating Debris)	2019	Solar-powered vessel with conveyor belt collects floating debris
	WasteShark	2016	Drone modeled after a whale shark skims the water and collects debris
	Jellyfishbot	2018	Remote-controlled robot collects garbage from waterways
	Seabin	2013	Automated bucket uses a pump to capture floating debris, including plastics
	BluePhin	2017	Battery-powered, zero carbon emissions robot uses artificial intelligence to collect floating waste
Boats and wheels	SeaVax	2015	Solar- and wind-powered ship collects plastics and other debris; sensors detect waste and sonar protects fish and other sea animals from being collected
	Mighty Tidy	2003	Trash skimming boat scoops plastics from the surface, and conveyor belt moves the waste to bin
	Inner-Harbor Water Wheel	2014	Wheel collects trash in the river before it enters into a harbor
	Versi-Cat Trash Skimmer Boat	2009	Skimmer collects floating and semi-submerged debris in a removable basket for later disposal
	ERVIS	2016	Ship with saucers uses centripetal force to capture and separate waste into five size classes for later disposal
	One-Earth One Ocean SeeHamster	2012	Small catamarans (4 × 12 m) equipped with fold-down nets or fishing gear collect debris from inland waters
	One-Earth One Ocean SeeKuh	2016	Nets with 2.5 cm mesh are suspended from catamarans (12 × 10 m) and collect plastic waste up to 4 m deep from bays, estuaries, and coastal regions
	Manta	2016	Ship brings waste on board for manual sorting and mechanical compacting before being carried to land for processing
	The Interceptor	2019	Solar-powered catamaran autonomously extracts floating plastic from rivers, using barriers and a conveyor belt
	MariClean	2020	Catamaran fitted with a conveyor collects debris from seas, straits, and bays

Prevention technology	Name of the technology	Year of invention	Process/Functioning
Detection aids	Malolo I	2017	Unnamed aerial robot detects marine debris (especially fishing gear) in the open ocean for later collection and satellite tagging
	Unnamed GPS Device on Ghost Nets	2019	Vessels place GPS on ghost nets to mark them for collection
	NetTag	2019	Low-cost transponders allow fishers to locate and recover lost nets
	Wikilimo	2019	Uses satellite imagery to detect major garbage patches in oceans; uses numerical models and machine learning to identify optimum routes for cleaning up garbage patches
Waterway litter traps	Bandalong Litter Trap	2009	Floating device uses waterway currents to capture and guide litter into the trap before it flows downstream
	Clear River Litter Trap	2014	Floating device uses waterway currents to capture and guide litter into the trap before it flows downstream
	SCG Litter Trap	2019	Floating litter trap uses a bypass flap to leverage water flow and pressure to capture and trap floating litter
River booms	Clean River Project River Boom	2005	Floating booms create a barrier that collects surface debris along the rivers
	Bandalong Boom	2015	Floating boom couplings span waterways to capture waste and prevent it from traveling further downstream
	The Litterboom Project	2017	Large pipes anchored across rivers catch surface-level debris
	AlphaMERS Floating Barrier	2015	Floating barricade carries debris to the riverbank for manual or mechanical collection
	Plastic Fischer Trap Boom	2019	Boom made of PVC pipe floaters and galvanized steel catching net collects surface plastics up to 60 cm deep
Sand filters	Barber Surf Rake	Not known	Tractor-towed machine removes waste on beaches
	Barber Sand Man	Not known	Walk-behind sand sifting machine uses a vibrating screen to sift debris from sand and soil on beaches
Microplastics			
Miscellaneous capture	Unnamed Invention by Anna Du	2018	Remotely operative vehicle uses infrared light to detect, photograph, and help remove microplastics from waterways
	Unnamed Invention by Fionn Ferreira	2019	Combination of oil and magnetite powder binds microplastics for extraction with a magnet
Sand filter	Marine Microplastic Removal Tool	2013	Sand is piled on a sheet of fine mesh stretched between two long poles, and the mesh catches plastic and other foreign material while allowing the sand to fall through

(Continued)

TABLE 11.2
(Continued)

Prevention technology	Name of the technology	Year of invention	Process/Functioning
Macro- and Microplastics			
Boat	OC-Tech	2013	Boat collects oil, microplastics, and other debris using a system of nets and baskets, clean water then flows back into ocean
Skimmer	Marina Trash Skimmer	2016	Pump in a partially submerged plastic box draws in and catches surface trash
Vacuum	Hoola One	2019	Vacuums approximately three gallons of sand and debris per minute into a tank that separates particles by buoyancy, allowing for plastic removal and separation
Air Barrier	The Great Bubble Barrier	2019	Tubes placed diagonally across the bottom of the waterway create a bubble barrier by pumping air, creating a current that brings debris to the surface and guides it to a catchment system.

Source: Based on the inventory made by Schmaltz et al. (2020).

Prevention technologies targeting macroplastics and microplastics, almost equal in number, are fewer in number as compared to those designed for collection purposes. Again, there are significantly more collection technologies targeting macroplastics than the ones targeting microplastics collection. Of course, capturing macroplastics before they finally enter into the ocean ecosystem itself amounts to targeting microplastics on a long-term basis. The plastic pollution collection depends more on the "boats and wheels" type technologies and the prevention of plastic pollution depends on the "storm and wastewater filters" type technologies (Tables 11.1, 11.2; Figures 11.2, 11.3).

Sometimes the microplastics are not mixed with plastic particles and are not easily identified. The Group of Experts on the Scientific Aspects of Marine Environmental Protection suggested that if the microplastics were divided into primary and secondary particles, it becomes easier to trace their source and identify remediation to decrease their inputs in the environment (GESAMP 2015). In case the microplastics bear distinctive shapes and colors, it might be possible to recognize them and, thus, to determine their sources; for example, microbeads in cosmetics sourced in primary microplastics and textile fibers as the secondary microplastics (Talvitie et al. 2017). An understanding about the microplastic source provides the possibility of controlling the microplastic contamination prior to their entry into WWTPs as well as in the environment. For instance, governments had begun regulating and imposing bans on microbeads usage in personal care items (Microbead-Free Waters Act 2015). In the same manner, filters for the removal of textile fibers from washing machines were designed (LIFE—MERMAIDS 2017). However, despite the removal of microplastics entering into WWTPs following the solutions based on the sources, a considerable proportion of unrecognized primary and secondary microplastics are still left, and this problem needs to be sorted out in the technologies to be developed in the future.

Microplastics widely occur in contaminated aquatic environments. When processed through WWTPs, their concentration in the influent has been recorded between 0 and

1047 molecules cc^{-1} (Priya et al. 2022). A study on the performance of treatment technologies for the removal of microplastics reveals that many operational technologies are efficient enough to remove microplastic pollution up to nearly 100% (Bui et al. 2020). Out of the nine technologies prevalent worldwide, the most efficient in performance recorded is that of the membrane bioreactor (MBR). Grit chamber/primary sedimentation, and coagulation treatment technologies can remove microplastics to an appreciable extent, up to 99%. Membrane disc filters and sand filtration have been found with microplastic removal performance up to 98.5 and 97%, respectively. The performance of dissolved air floatation (DAF), conventional activated sludge/secondary sedimentation, and ozonation technologies was recorded as 95, 93.8, and 90% respectively. The performance of granular activated carbon (GAC), as recorded in China, is comparatively poor, ranging between 56.8 to 60.9% (Table 11.3). The figures in the table show the highest recorded performance. In many cases, as documented by Bui et al. (2020), the performance of each of the technologies has also been found much less than the potential performance indicated in the table. The performance of the treatment technologies, in fact, depends on their operating conditions.

TABLE 11.3
Performance of treatment technologies for microplastic removal.

Technology	Operating conditions	Removal (%)	References
Grit chamber/Primary sedimentation	Passing through a 6-mm screen mesh before grit chamber	99	Lares et al. (2018)
Dissolved air floatation—DAF	Flocculent: PAC of 40 mg L^{-1}	95	Talvitie et al. (2017)
Coagulation	Microplastics with size >10 μm	>99	Wang et al. (2020)
Sand filtration	Filtration bed: 1 m of gravel (3–5 mm size) + 0.5 m of quartz (0.1–0.5 mm size)	97	Talvitie et al. (2017)
Granular activated carbon—GAC	Coagulants: PAC 40 mg L^{-1}, PAM 0.001–0.002 mg L^{-1}; pH of 7.70–7.84	56.8–60.9	Wang et al. (2020)
Membrane discfilter	Hydrotech HSF 1702–1F, two discs with 24 filter panels for each disc; HRT: 4 min; filtration area: 5.76 m^2; Fe-based coagulant: 2 mg L^{-1}; cationic polymer: 1 mg L^{-1}; media (pore size: 20 μm)	98.5	Talvitie et al. (2017)
Conventional activated sludge/secondary sedimentation	SRT: 28 3days; HRT: 4–8 h; pH: 6.3–7.3; MLSS: 3100–4200 mg L^{-1}; temperature: 8–18°C	93.8	Lares et al. (2018)
Membrane bioreactor—MBR	Submerged MBR (KUBOTA flat-sheet; HRT: 20–100h)	100	Talvitie et al. (2017)
Ozonation	O$_3$ contact time: 60 min: temperature: 35–45°C	>90	Chen et al. (2018)

Note: HRT—hydraulic retention time; MLSS—mixed liquor suspended solids; PAC—polyaluminum chloride; PAM—polyacrylamide; SRT—sludge retention time.

While many technologies addressing macro- and microplastic pollution through prevention and removal have been found performing up to mark, there is no point of complacency, as no technology can be an ultimate panacea until and unless very sound and foolproof measures combined with the most efficient technologies are strictly implemented at a global scale. No single technology, in fact, can be claimed to be the final one. As the technology is continuously in evolution, more efficient and worthwhile replicable technologies will certainly be developed and widely adopted in the future. The sampling methods of microplastics as well as their detection and quantification methodologies are somewhat varying. For the appropriate results on the performance of the removal and prevention technologies, it would be better to harmonize these methodologies. The source monitoring should ensure removal of the microplastic pollutants which is of absolute importance for avoiding a chain of contaminations. Separation of microplastics from wastewater to the extent of 100% at household, manufacturing industry, and recycling plant levels will be the future endeavors of addressing the problem.

11.3 SUMMARY

The cutting-edge type solutions have in their folds the technologies developed in recent years that are appropriate for targeting marine environments contaminated with plastic pollutants: the prevention and collection/cleaning-up technologies. There is stray leakage of plastic particles at every stage of the plastic life cycle: pre-production (in the form of pellets), production, consumption, and waste management (recycling; reuse/repurposing, such as waste-to-energy; and landfilling). Innovative technologies are applicable at various stages of the plastic cycle at which plastic pollutants' leakage is occurring, to prevent or minimize the leakage and contribute to averting or reducing microplastic pollution that eventually becomes marine pollution. As about 80% of the plastic pollution in the marine ecosystem is from land-based sources, efficient prevention technologies applied can be instrumental in reducing microplastics load in the marine ecosystem.

Schmaltz and co-workers have documented 52 types of prevention and collection technologies targeting macro- and microplastics that are currently in use in the world. The prevention technologies are 14 in number. Among the 38 collection technologies, 31 are largely applicable for macroplastics, three exclusively for microplastics and four are applied for both macro- and microplastics. Prevention technologies targeting macroplastics and microplastics, almost equal in number, are fewer in number as compared to those designed for collection purposes. Again, collection technologies meant for macroplastics are significantly larger in number than the ones targeting microplastics. The plastic pollution collection depends more on the "boats and wheels" type technologies and the prevention of plastic leakage on the "storm and wastewater filters" type technologies. These technologies are apart from the prevailing membrane technologies of tackling microplastic pollutants.

An understanding about the microplastic source provides the possibility of controlling the microplastic contamination prior to their entry into WWTPs as well as into the environment. Many operational membrane technologies are efficient enough to remove microplastic pollution up to nearly 100%. Out of the nine technologies prevalent worldwide, the most efficient in performance recorded is that of the membrane bioreactor (MBR). Grit chamber/primary sedimentation, coagulation treatment, membrane discfilter, sand filtration, dissolved air floatation (DAF), conventional activated sludge/secondary sedimentation, and ozonation technologies also perform very well.

No single technology, in fact, can be claimed to be the ultimate one. As the technology is continuously in evolution, more efficient and worthwhile replicable technologies will certainly be designed, developed, and widely adopted in the future.

REFERENCES

Bui, X-T., Vo, T.-D.-H., Nguyen, P.-T., Nguyen, V.-T., Dao, T.-S. and Nguyen, P.-D. 2020. Microplastic pollution in wastewater: Characteristics, occurrence and removal technologies. *Environmental Technology and Innovations*, 19:101013. doi: 10.1016/j.eti.2020.101013.

Chen, R., Qi, M., Zhang, G. and Y. C. 2018. Comparative experiments on polymer degradation technique of produced water of polymer flooding oilfield. In: *IOP Conference Series: Earth and Environmental Science*. IOP Publishing. 012208. doi: 10.1088/1755-1315/113/1/012208.

GESAMP (Group of Experts on the Scientific Aspects of Marine Environmental Protection). 2015. Sources, fate and effects of microplastics in the marine environment: A global assessment. In: Kershaw, P. J. (ed.), *IMO/FAO/UNESCO-IOC/UNIDO/WMO/UN/UNDP Joint Group of Experts on the Scientific Aspects of Marine Environmental Protection*. London: Rep. Stud. GESAMP No. 90. 96.

Lares, M., Ncibi, M. C., Sillanpää, M. and Sillanpää, M. 2018. Occurrence, identification and removal of microplastic particles and fibers in conventional activated sludge process and advanced MBR technology. *Water Research*, 133:236–246. doi: 10.1016/j.watres.2018.01.049.

Lau, W. W. Y., Shiran, Y., Bailey, R. M., Cook, E., Stuchtey, M. R., Koskella, J., Velis, C. A., Godfrey, L., Boucher, J., Murphy, M. B., Thompson, R. C., Jankowska, E., Castillo, A. C., Pilditch, T. D., Dixon, B., Koerselman, L., Kosior, E., Favoino, E., Gutberlet, J., Baulch, S., Atreya, M. E., Fischer, D., He, K. K., Petit, M. M., Sumaila, U. R., Neil, E., Bernhofen, M. V., Lawrence, K. and Palardy, J. E. 2020. Evaluating scenarios towards zero plastic pollution. *Science*, 369(6510):1455–1461. doi: 10.1126/science.aba9475.

Microbead-Free Waters Act. 2015. https://obamawhitehouse.archives.gov/the-press-office/2015/12/28/statement-press-secretary-hr-1321-s-2425.

Priya, A., Anusha, G., Thanigaive, S., Karthick, A., Mohanave, V., Velmurugan, P., Balasubramanian, B., Ravichandran, M., Kamyab, H., Kirpichnikova, I. M. and Cheliappan, S. 2022. Removing microplastics from wastewater using leading-edge treatment technology: A solution to microplastic pollution—A review. *Bioprocess and Biosystems Engineering*. doi: 10.1007/s00449-022-02715-x.

Ritchie, H. and Roser, M. 2018. Plastic pollution: Our world in data. https://ourworldindata.org/plastic-pollution.

Schmaltz, E., Melvin, E. C., Diana, Z., Gunady, E. F., Ritsschof, D., Somarelli, J. A., Virdin, J., Dunphy-Daly, M. M. 2020. Plastic pollution solutions: Emerging technologies to prevent and collect marine plastic pollution. *Environment International*, 144:106067. doi: 10.1016/j.envint.2020.106067.

Talvitie, J., Mikola, A., Koistinen, A. and Setal, O. 2017. Solutions to microplastic pollution—Removal of microplastics from wastewater effluent with advances wastewater treatment technologies. *Water Research*, 123:401–407. doi: 10.1016/j.watres.2017.07.005

United Nations Environment Assembly. 2016. Resolution 2/11. Marine plastic litter and microplastics. http://nicholasinstitute.duke.edu/plastics-policies/unea-resolution-211-marine-plastic-litter-and-microplastics.

Wang, Z., Lin, T. and Chen, W. 2020. Occurrence and removal of microplastics in an advanced drinking water treatment plant (ADWTP). *Science of Total Environment*, 700:134520. doi: 10.1016/j.scitotenv.2019.134520.

World Economic Forum. 2016. *The New Plastic Economy: Rethinking the Future of Plastics.* Geneva. 34pp. https://www3.weforum.org/docs/WEF_The_New_Plastics_Economy.pdf.

Worm, B., Lotze, H. K., Jubinville, I., Wilcox, C. and Jambeck, J. 2017. Plastic as a persistent marine pollutant. *Annual Review of Environment and Resources*, 42(1):1–26. doi: 10.1146/annurev-environ-102016-060700.

Zheng, J. and Suh, S. 2019. Strategies to reduce the global carbon footprint of plastics. *Nature Climate Change*, 9:374–378. doi: 10.1038/s41558-019-0459-z.

Index

Note: Page numbers in *italics* indicate a figure and page numbers in **bold** indicate a table on the corresponding page.

For Product Safety Concerns and Information please contact our EU
representative GPSR@taylorandfrancis.com
Taylor & Francis Verlag GmbH, Kaufingerstraße 24, 80331 München, Germany

www.ingramcontent.com/pod-product-compliance
Lightning Source LLC
Chambersburg PA
CBHW060443240326
41598CB00087B/3075